Selected Exercises for the Biochemistry Laboratory

SELECTED EXERCISES FOR THE
BIOCHEMISTRY LABORATORY

G. DOUGLAS CRANDALL

Emmanuel College

New York   Oxford   •   OXFORD UNIVERSITY PRESS   •   1983

ISBN O-19-5O3185-7

Printing (last digit): 9 8 7 6 5 4 3 2 1

Printed in the United States of America

To my son, Chris

PREFACE

The past four decades have witnessed unprecedented advances in the field of biochemistry. As a result, our present understanding of biological systems at the molecular level is well beyond that ever anticipated by the biochemists of the 1940's and 1950's. This impressive record of progress has in no small part been due to the existence of an entire array of imaginative laboratory techniques and procedures. These are the tools of the biochemist and permit precise analysis of the numerous biomolecules of the cell and a determination of their respective cellular roles. It is therefore of utmost importance for those of us involved in the education of biochemists to provide the best possible training in the laboratory as well as the classroom. Our aim in this should be twofold, first to teach the basic skills necessary to work sucessfully in a modern biochemistry laboratory and second to provide a learning experience that instills students with the confidence to work independently in the laboratory. The exercises selected for this laboratory manual are designed to accomplish both goals.

This manual is intended for use in a one-semester introductory biochemistry course. In addition to the laboratory manual, each student should have and maintain a laboratory notebook. The manual provides all the necessary instructions for each exercise plus notes and cautions highlighting more difficult or potentially hazardous operations. The notebook should be used as a detailed record of what actually occurs in the laboratory, including numerical data, graphs, drawings, and calculations. The questions at the end of each exercise aid in evaluating data and drawing conclusions, and the answers to these should also be entered in the notebook. In my own experience the notebook (or carbon copies thereof) serves as an excellent basis for grading student progress.

The most unique feature of this laboratory manual is the manner in which the level of difficulty and degree of independence increases gradually throughout the course of the semester. The first two chapters are relatively simple exercises introducing standard techniques that are used throughout the course. Tables, graphs, and complete instructions explaining data collection and treatment are included. Students with a minimum of chemistry background should be able to complete these exercises with a high degree of success. Subsequent chapters include exercises that illustrate the properties of major groups of biomolecules and introduce new techniques. These chapters are more detailed, require more time, and provide an ever-increasing challenge. The trend toward independence is culminated in the final chapter, which replaces the step-by-step protocol with an original journal article. Students, guided by a series of questions, are expected to use this article to prepare their own detailed procedure for the isolation of DNA and to implement the procedure in the laboratory.

Since the work described in most chapters requires two or more successive laboratory periods there are more exercises than can reasonably be completed in a single academic semester. This is by design and allows instructors to select those exercises which best fit their needs and interests. The manual therefore may be used in its entirety or selectively, depending on the sophistication of the students and the pedagogical needs of the instructor. This versatility commends its use in a number of different kinds of biochemistry courses.

The appendix is a detailed set of instructions for the preparations necessary to carry out each exercise. The amounts of equipment and reagents suggested assume a group of ten students working together in pairs. The instructions are sufficiently detailed that anyone with a minimum of scientific background will be able to make the preparations easily.

As in the case of any work of this nature, there are many people who deserve thanks for their contributions. The typists who worked to transform my long-hand manuscript and subsequent drafts into its present form include Carol Rosaly-Campos, Antonia Rosario, Ginette Garcia, Sharon Kinny, Anne Marie St. Pierre, and Niki Eastman-Eich. Special thanks are also due to William L. Miller of North Carolina State University, who read the entire manuscript and offered many helpful recommendations for its alteration, and to Julius Marmur of the Albert Einstein School of Medicine for permission to reprint his classic paper on DNA isolation. I am deeply indebted to Michael R. Cook, my editor at Oxford, who orchestrated the entire project with skill and patience, and thanks are also due to all of the students in my course, *Selected Topics in Biochemistry,* whose comments, criticism, and advice have made this book a thoroughly field-tested product. My most profound gratitude is reserved for my wife, Jane, who served as typist, proofreader, and editor and in many other roles.

G. D. C.

Boston

CONTENTS

Selected Exercises for the Biochemistry Laboratory

1 BASIC TECHNIQUES: THE PREPARATION OF AQUEOUS SOLUTIONS FOR THE LABORATORY

The living cell is composed primarily of water (80 to 90 percent by weight), and practically all of the chemical reactions typical of living systems occur in aqueous solution. To study these reactions in an orderly fashion outside the cell *(in vitro)* and obtain meaningful information about their activity inside the cell *(in vivo),* the biochemist must be able to prepare media that mimic the aqueous milieu of the cell. Two parameters of the solution that must be accurately known and strictly maintained are the concentration of critical salts and the concentration of hydrogen ions.

Concentration of Critical Salts

Concentration is the quantitative expression of the amount of a given solute dissolved in a solvent. Several methods have been developed for the preparation of solutions, and two of those are especially useful to biochemists: percent solution and molarity.

The per cent solution is useful mainly because of its ease of preparation. The term *per cent* means *parts per hundred;* thus one might logically conclude that a 10 per cent solution of sodium chloride contains 10.0 g of NaCl and 90.0 g of water. This would be a correct assumption as long as the solution was prepared by the weight per weight method, which is but one of three procedures for preparing per cent solutions. Such a solution should be labeled 10% NaCl (w:w) to indicate that it was prepared in this manner. The second alternative, termed the weight per volume method, calls for the dissolution of 10.0 g of NaCl in a sufficient quantity of distilled water to give a final volume of 100.0 ml.

This solution should be labeled 10% NaCl (w:v). As one can readily surmise, the final concentrations of these two preparations are quite similar, but not identical. It is therefore important that the solutions be accurately labeled.

The third method for preparing a per cent solution is often used when both the solvent and the solute are liquids, as in the case of ethanol dissolved in water. A 25 per cent solution of ethanol (v:v) is prepared by adding 25.0 ml of ethanol to 75.0 ml of distilled water.

Molar concentrations do not suffer from the potential inaccuracies of per cent solutions because they are based upon the number of molecules dissolved in the solvent. A molar solution is prepared by dissolving 1 gram molecular weight* of a substance in 1 liter of solution. Since a gram molecular weight, also called a mole, contains 6×10^{23} molecules, a molar solution of any substance contains 6×10^{23} molecules per liter.

Thus a 1 molar solution of sodium chloride (1.0 *M* NaCl) is prepared by dissolving the number of grams of NaCl equal to its molecular weight (58.0) in a sufficient quantity of solvent to give a final volume of 1 liter of solution. In a similar fashion 1.0 *M* NaOH (M.W. = 40.0) is prepared by dissolving 40.0 g of NaOH in a final volume of 1000 ml. It is important to note that although different weights of each substance are used to prepare these two solutions, they are equimolar and contain exactly the same number of molecules per given volume.

---

* 1 gram molecular weight = the number of grams equal to the molecular weight.

Other units of concentration based upon molarity are indicated in Table 1-1. You should familiarize yourself with these units because most substances occur at extremely low concentrations in the cell.

Hydrogen Ion Concentration

Many biochemical reactions are sensitive to the hydrogen ion concentration $[H^+]$; therefore the ability to monitor and control this parameter is important. The term *pH* is used to express hydrogen ion concentration and is calculated from Sorensen's equation:

$$pH = -\log_{10}[H^+]$$

Solutions with a pH equal to 7 (Table 1-2) are called neutral because the concentration of hydrogen ions $[H^+]$ is equal to that of the hydroxyl ions $[OH^-]$. Those solutions possessing high levels of $[H^+]$ are considered acidic (pH < 7), and those with low $[H^+]$ are basic (pH

> 7). The pH scale is illustrated in Table 1-2. As is evident from this table, natural fluids such as blood and lemon juice may have very different hydrogen ion concentrations. It is important to bear in mind that the pH scale is a logarithmic one and a difference of one pH unit represents a tenfold difference in hydrogen ion concentration. Consequently the difference in hydrogen ion concentration between lemon juice (pH = 2.3) and blood plasma (pH = 7.4) is approximately five orders of magnitude, or 100,000-fold.

Measurement of pH

Certain dye substances in solution exhibit different colors according to the pH. For this reason, they are termed indicator dyes and are often used to determine pH values. This colorimetric technique is not as accurate as the electrometric method, but there are instances when it is useful and even preferred. Note in Table 1-3 the pH range and the color changes at the extremes of the range for each indicator.

A modification of the colorimetric method uses paper impregnated with a combination of indicator dyes. The application of a liquid sample to the paper causes a color change characteristic of the pH of the sample. In practice, one applies a drop of the sample to the pH indicator paper and compares the resultant color with a standard color key. This method, although crude, has the advantages of speed and convenience.

The most accurate pH measurements are done electrometrically using a pH meter. This instrument is composed of an electrode connected to an electrometer which records very small differences in electric potential

Table 1-1.
Units of concentration

Name	Abbreviation	Fraction of a molar solution
Molar	M	1
Millimolar	mM	10^{-3}
Micromolar	μM	10^{-6}

Table 1-2.
The pH scale

$[H^+]$ (M)	pH	Examples
10^0	0	
10^{-1}	1	0.1 M HCl
10^{-2}	2	Lemon juice
10^{-3}	3	Gastric juice
10^{-4}	4	Tomato juice
10^{-5}	5	Urine
10^{-6}	6	
10^{-7}	7	Pure water, blood plasma, seawater
10^{-8}	8	
10^{-9}	9	
10^{-10}	10	
10^{-11}	11	
10^{-12}	12	0.01 M NaOH
10^{-13}	13	
10^{-14}	14	

Table 1-3.
Some common colorimetric indicators and their effective ranges

Indicator	pH range	Acid color	Base color
Thymol blue	1.2–2.8	Red	Yellow
Bromophenol blue	3.0–4.6	Yellow	Blue
Bromocresol green	3.8–5.4	Yellow	Blue
Methyl red	4.2–6.3	Red	Yellow
Paranitrophenol	6.2–7.5	Colorless	Yellow
Cresol red	7.2–8.8	Yellow	Red
Phenolphthalein	8.0–9.8	Colorless	Red

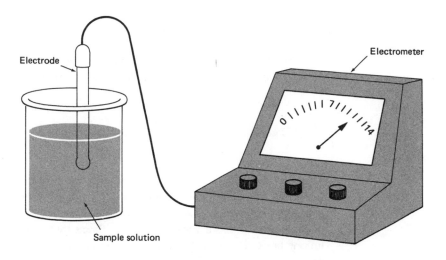

Fig. 1-1. Diagram of a pH meter.

(Fig. 1-1). The tip of the electrode is immersed in the sample, and the pH is registered on the electrometer dial.

Buffered Solutions

The pH of the cell is generally maintained at or near neutrality because most enzymes function best at this pH. This is accomplished by the presence of weak acids, such as carbonic and phosphoric, which with their conjugate bases act as buffers. A buffered solution is defined as one that resists pH change upon the addition of acid or base. Such solutions are used in many biochemical experiments.

The acetate buffer system consists of a mixture of acetic acid (a weak acid) and sodium acetate (its conjugate base):

$$CH_3COOH \rightleftharpoons CH_3COO^- + H^+$$
$$CH_3COONa \rightleftharpoons CH_3COO^- + Na^+$$

If an acid is added to such a solution, the excess hydrogen ions are neutralized by combination with the acetate ions, thus maintaining the original pH:

$$CH_3COOH \rightleftharpoons CH_3COO^- + H^+$$
$$\text{Added } H^+$$
$$CH_3COOH$$

If a base is added to the acetate buffer system, the excess hydroxyl ions are neutralized by the hydrogen ions, again maintaining the original pH:

$$CH_3COOH \rightleftharpoons CH_3COO^- + H^+$$
$$\text{Added } OH^-$$
$$H_2O$$

Careful titration of acid or base into the acetate system (Figure 1-2) demonstrates that the range of greatest buffering capacity is limited to about one pH unit either side of the pK, which is the negative logarithm of the dissociation constant of the acid. The pK for acetic acid is 4.76; thus the acetate buffer is most effective from pH 3.76 to pH 5.76.

The titration curve shown in Figure 1-2, which is common to most weak organic acids, is expressed mathematically by the Henderson-Hasselbalch equation:

$$pH = pK + \log_{10} \frac{[\text{conjugate base}]}{[\text{acid}]}$$

This equation is extremely useful because it can be used to prepare different buffer systems which vary with regard to pH and buffering capacity. Table 1-4 illustrates only a few of the wide variety of buffer systems available to the biochemist; it is also interesting to notice

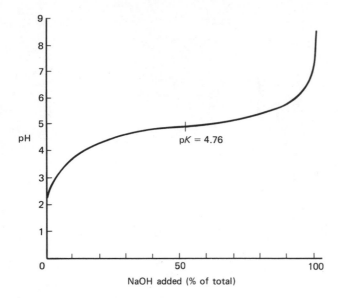

Fig. 1-2. Titration of a strong base (NaOH) against a weak organic acid.

that some acids have two or three pK values and can act as buffers in several pH ranges. For example, malic acid could be used to prepare a solution with buffering capacity in a pH range of 2.4 to 4.4 or 4.26 to 6.26.

If you want to prepare a solution that buffers well at pH 3, formic acid would be an appropriate choice. Substitution of the desired pH (3.0) and the pK for formic acid (3.75) in the Henderson-Hasselbalch equation permits the calculation of the necessary concentrations of formic acid and formate. As you might guess, the Henderson-Hasselbalch equation is used widely in the preparation of buffers for laboratory use.

Table 1-4.
Some weak acids and their corresponding pK values

Weak acid	pK_1	pK_2	pK_3
Formic	3.75		
Acetic	4.76		
Lactic	3.86		
Malic	3.40	5.26	
Carbonic	6.35	10.30	
Citric	3.09	4.75	5.41
Phosphoric	2.12	7.21	12.66

The following exercises are sufficiently simple that if you have done the calculations in advance, you will have time to check them over with the laboratory instructor and still be able to complete them within a 3 to 4 hour period.

Basic Techniques I. Solution Preparation

Introduction

This exercise serves to familiarize you with some of the basic calculations that must be done prior to solution preparation and with the fundamental laboratory skills necessary to prepare those solutions. Since these solutions are to be used in subsequent exercises, it is important that you work as carefully as possible. Your task is to prepare the following five solutions:

0.1 *M* sodium acetate

0.1 *M* hydrochloric acid

0.1 *M* sodium hydroxide

0.1 *M* acetic acid

0.1 *M* acetate buffer (pH 5.0)

You will be given instructions for the preparation of the sodium acetate and HCl solutions and are expected to use those guidelines for the preparation of the others.

Procedure

1. Prepare 1 liter of 0.1 *M* sodium acetate (M.W. = 136.08).
   a. Calculations
      1) One gram molecular weight of sodium acetate is 136.08 g.
      2) A 0.1 *M* solution contains 0.1 gram molecular weight per liter (0.1 × 136.08 = 13.61 g/liter).
   b. Preparation
      1) Weigh out 13.61 g of sodium acetate on a triple-beam balance.
      2) Transfer the reagent to a 1-liter volumetric flask and add approximately 500 ml of distilled water.
      3) Dissolve the powder by swirling and then

add sufficient distilled water to bring the volume up to exactly 1 liter.

4) Transfer the solution to a clean, dry flask and label it 0.1 *M* sodium acetate.

NOTE: Be sure to rinse the volumetric flask after each solution preparation.

2. Prepare 1 liter of 0.1 *M* hydrochloric acid (M.W. = 36.5).

   a. Calculations

   The calculations necessary for this solution are slightly more involved because two other factors must be taken into consideration: the purity and the density of the concentrated HCl from which the solution must be prepared. Most commercially available HCl is 37 per cent pure and has a density of 1.19 g/ml. Check the bottle of HCl you will be using to confirm this.

   1) One gram molecular weight of HCl is 36.5 g.
   2) A 0.1 *M* solution will contain 3.65 g of HCl per liter.
   3) What volume of the concentrated HCl contains exactly 3.65 g of HCl?
      a) 1.0 ml of concentrated HCl weighs 1.19 g and is 37 per cent pure. Thus, 1.0 ml of concentrated HCl contains 0.37×1.19 g = 0.44 g of pure HCl.
      b) If 1.0 ml contains 0.44 g of HCl, then the volume containing 3.65 g can be determined from the following proportionality:

      $$\frac{0.44 \text{ g}}{1.0 \text{ ml}} = \frac{3.65 \text{ g}}{x \text{ ml}}$$

      x = 8.29 ml (the volume of concentrated HCl containing exactly 3.65 g of pure HCl).

   b. Preparation

CAUTION: Concentrated hydrochloric acid is an extremely volatile and caustic reagent and can cause severe burns if not handled properly. Do not pipette this reagent by mouth! Concentrated hydrochloric acid should be transferred with a propipette under the fume hood.

   1) Place about 500 ml of distilled water in a 1-liter volumetric flask.

2) Use a 10-ml pipette to transfer 8.29 ml of concentrated HCl to the flask. The acid should be added slowly and dissolved by gentle swirling. Add distilled water until the volume is exactly 1 liter.
3) Store the solution in a labeled flask.

3. Prepare 1 liter of 0.1 *M* sodium hydroxide (M.W. = 40.0).
   a. Calculations (use this space for your calculations).

   b. Preparation (use this space to write out how you plan to prepare the solution).

NOTE: Sodium hydroxide pellets are very caustic and should be handled only with a spatula. The pellets are also quite hygroscopic and absorb atmospheric moisture very rapidly. Thus the bottle should not be left uncapped any longer than necessary.

4. Prepare 1 liter of 0.1 *M* acetic acid (M.W. = 60.1).
   a. Calculations (concentrated acetic acid, usually called glacial acetic, has a density of 1.05 g/ml and is 99.7 per cent pure)

   b. Preparation

CAUTION: Glacial acetic acid should be handled with the same precautions used for concentrated HCl.

5. Prepare a 0.1 *M* acetate buffer (pH 5.0) from the 0.1 *M* acetic acid and 0.1 *M* sodium acetate.
   a. Calculations
      1) Use the Henderson-Hasselbalch equation to determine the ratio of sodium acetate to acetic acid necessary to yield a pH of 5.0:

$$pH = pK + \log_{10} \frac{[\text{conjugate base}]}{[\text{acid}]}$$

$$5.0 = 4.76 + \log_{10} \frac{[\text{acetate}]}{[\text{acetic acid}]}$$

$$0.24 = \log_{10} \frac{[\text{acetate}]}{[\text{acetic acid}]}$$

$$\frac{[\text{acetate}]}{[\text{acetic acid}]} = 1.74$$

   b. Preparation
      1) Use a graduated cylinder to measure 174 ml of 0.1 *M* sodium acetate and then add it to 100 ml of 0.1 *M* acetic acid.
      2) Store the solution in a flask labeled 0.1 *M* acetate buffer (pH 5.0).

Basic Techniques II. Measurement of pH

Introduction

This is your opportunity to check the accuracy of your previous work by determining the pH of each solution using the colorimetric method and the electrometric method. By comparison of results, the precision of the two methods may be evaluated.

Procedure

1. Electrometric method (pH meter)
   a. Transfer about 10 ml of the 0.1 *M* acetic acid solution to a 20-ml beaker.
   b. Lower the pH electrode into the solution so that the tip of the electrode is totally immersed. Do this with care so as to avoid breaking the electrode on the bottom of the beaker.
   c. Adjust the pH meter to record pH to the nearest tenth of a pH unit. Instructions for the operation of pH meters vary from model to model, and so you should consult the laboratory instructor for precise directions.
   d. Record the pH of 0.1 *M* acetic acid in Table 1-5.

Table 1-5.
Electrometric and colorimetric determinations of pH

Solution (0.1 M)	pH reading	
	Meter	Paper
Acetic acid		
Sodium acetate		
Acetate buffer		
hydrochloric acid		
Sodium hydroxide		

e. Remove the electrode and rinse it with a steady stream of distilled water from a wash-bottle.

f. Repeat the same steps for all five solutions and record your data in Table 1-5.

2. Colorimetric method (pH indicator paper)

a. With a Pasteur pipette, place a single drop of the solution on a strip of pH paper (sensitivity range 0–13). *Do not* dip the pH paper into the solution because the dyes will leach out and contaminate the sample.

b. Immediately compare the color of the paper with the standard color key and decide which color it most nearly matches. Record the pH in Table 1-5.

c. Repeat the procedure for all five solutions and record the results in Table 1-5. Be sure to use clean glassware for each determination.

Calculations and Questions

1. Compare the pH values obtained by each method.

a. Are they identical? How much do they vary?

b. Would you feel comfortable using pH paper in the absence of a pH meter? Why?

2. Use Sorenson's equation ($pH = -\log_{10}[H^+]$) to calculate the pH of your 0.1 M HCl and 0.1 M NaOH solutions. You can assume that each of these is fully dissociated in aqueous solution.

a. How close are the calculated and measured pH values?

b. Investigate the literature (Cooper, 1977) to discover some of the limitations of pH measurement that may contribute to variation between theoretical and observed values.

3. Based upon your initial instructions, the acetate buffer should have a pH of 5.0.

a. How closely does the observed pH approach the theoretical pH?

b. If the measured pH varies by more than 0.1 pH unit, you should prepare the buffer again. Remember that the pH scale is logarithmic and that a relatively small variation in pH represents a much larger disparity in hydrogen ion concentration.

Basic Techniques III. Buffering Capacity

Introduction
Compare the buffering capacity of water with that of acetate buffer. This is done by adding small, measured amounts of base to each of these samples and recording the pH after each addition. As you might suspect, the results for water, which is not buffered at all, will be considerably different from those for the acetate solution.

Procedure
1. Determine to what extent the addition of strong base (0.1 M NaOH) changes the pH of water.

a. Pipette 10.0 ml of distilled water into a 50-ml flask.

b. Determine the pH with pH paper and record the value in Table 1-6.

c. Using a clean Pasteur pipette, add 2 drops of 0.1 M NaOH to the water, mix thoroughly, and measure the pH again.

d. Continue in this fashion—adding 2 drops of 0.1 M NaOH to the sample at a time, measuring the pH colorimetrically, and recording the data—until approximately 20 drops have been added (or until the pH has risen to about 11–13).

Table 1-6.

pH changes in water as strong base is added

Total amount of 0.1 M NaOH added	pH
O drops	
2	
4	
6	
8	
1O	
12	
14	
16	
18	
20	

Table 1-7.

pH changes in acetate buffer as strong base is added

Total amount of 0.1 M NaOH added	pH
O drops	
2	
4	
6	
8	
10	
12	
14	
16	
18	
20	

2. Determine to what extent the addition of strong base (O.1 M NaOH) changes the pH of O.1 M acetate buffer (pH 5.O).
   a. Transfer 1O.O ml of O.1 M acetate buffer (pH 5.O) to a 5O-ml flask and determine the pH.
   b. Make 2-drop additions of O.1 M NaOH and record the pH after each addition until you have added the same number of drops to the acetate buffer as you added to the water in the previous exercise.
   c. Record these data in Table 1-7.

Calculations and Questions

1. Plot the pH data from Tables 1-6 and 1-7 on the graph paper provided (Figure 1-3). Be sure to label both plots carefully.
2. To what do you ascribe the differences in pH change between water and acetate buffer?
3. What do you think the graphs might have looked like if you had added strong acid (HCl) to the water

and acetate buffer? Prepare a graph to illustrate your answer.

Basic Techniques IV.
pH and Buffering Capacity
of Some Natural Solutions

This part of the exercise is optional and gives you an opportunity to experiment with the techniques you have mastered. The laboratory instructor will have some solutions of natural origin (tea, vinegar, lemon juice, blood, etc.) which you should test for pH and buffering capacity. Think of some other similar liquids that might also provide interesting experimental information.

Additional Reading

F. B. Armstrong, *Biochemistry,* 2nd ed., Chap. 3. Oxford University Press, New York, 1983.

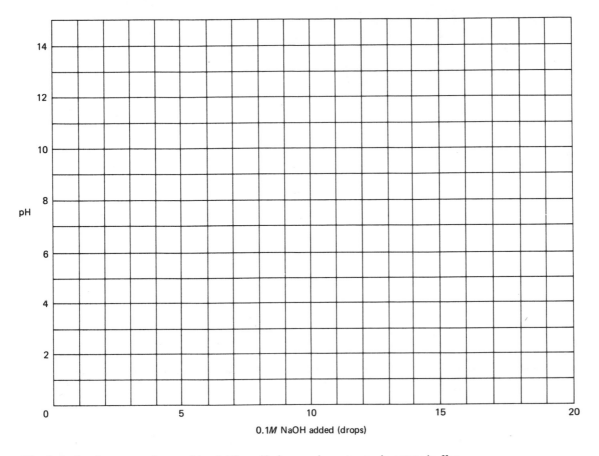

Fig. 1-3. Graph paper to be used in plotting pH changes in water and acetate buffer.

T. G. Cooper, *The Tools of Biochemistry,* Chap. 1. Wiley, New York, 1977.

A. L. Lehninger, *Biochemistry,* 2nd ed. Worth, New York, 1975.

J. G. Morris, *A Biologist's Physical Chemistry.* Addison-Wesley, Reading, MA, 1968.

G. Rendina, *Experimental Methods in Modern Biochemistry.* Saunders, Philadelphia, 1971.

H. A. Sober and R. A. Harte, eds., *Handbook of Biochemistry, Selected Data for Molecular Biology.* Chemical Rubber, Cleveland, 1968.

D. C. Wharton and R. E. McCarty, *Experiments and Methods in Biochemistry.* Macmillan, New York, 1972.

2 COLORIMETRY: A SPECTROPHOTOMETRIC ANALYSIS OF RIBOFLAVIN

Colorimetry

Colorimetry is one of the most widely used techniques in the biochemistry laboratory. This technique can provide both quantitative and qualitative information regarding dissolved substances. The colorimeter is an instrument designed to direct a beam of parallel, monochromatic light through a liquid sample and measure the amount of light that emerges. If the sample is pure solvent (e.g., water), the ratio of transmitted light intensity (I) to incident light intensity (I_o) is equal to unity. This value is referred to as transmittance (T) and is expressed as a percentage:

$$\% T = \frac{I}{I_o}$$

When a colored solution is exposed to visible light, some portion of the incident light is absorbed by the solution and thus the transmitted light is reduced. For instance, a solution that absorbs one quarter of the incident light has a transmittance of 75 per cent. Other factors—for example, light scattering due to the presence of suspended particles—may also reduce the percentage of light transmitted. Figure 2-1 illustrates the fundamental internal components of a simple colorimeter.

The Lambert-Beer Law

A more precise expression of the relationship between the amount of light absorbed and the concentration of a solution is the combined Lambert-Beer law:

$$A = Ec\ell$$

where: A = absorbance
E = molar extinction coefficient (liter mole^{-1} cm^{-1})
c = concentration (moles/liter)
ℓ = optical path length (cm)

Thus, the absorbance is equal to the product of the molar extinction coefficient, the concentration of the solution, and the distance traversed by the light beam. The molar extinction coefficient is a constant for a molecular species dissolved in a given solvent and measured at a specified wavelength. It should be noted that the term *optical density* (O.D.) is often used synonymously with *absorbance* although they are not precisely interchangeable. *Absorbance* is the preferred term and refers to reduction in transmitted light caused strictly by absorption, as occurs in dilute solutions. If the light reduction can be attributed to light scattering as well, the term *optical density* should be used. This is a source of some confusion because most colorimeters are calibrated in O.D.

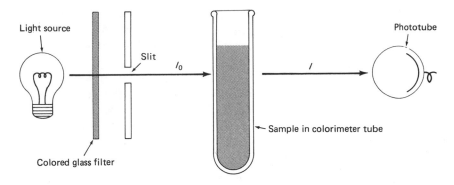

Fig. 2-1. A diagram of the fundamental internal components of a colorimeter. The filter selects a specific wavelength of light, and the slit makes the light parallel. Light transmitted by the sample (I) is picked up by the phototube, which is connected to a dial or a digital readout.

Applications of Colorimetry

The obvious application of the Lambert-Beer law is the use of the colorimeter to determine the concentration of a wide variety of light-absorbing molecules, such as carotene, chlorophyll, and hemoglobin. For example, three of the variables needed to determine the concentration of a carotene solution (E, A, and ℓ) can be obtained by looking up the molar extinction coefficient of carotene and empiracally determining the absorbance and the path length. It is then a simple matter to solve for the remaining variable, c. With a little ingenuity the technique can be extended to measure the concentration of colorless substances, such as sugars and amino acids. This requires the reaction of the colorless moiety with some other atomic or molecular species to form a soluble, colored complex. The concentration of the complex can then be determined as previously described.

The absorption spectrum, a unique pattern of absorption maxima and minima at different wavelengths, is often used in biochemistry to characterize or identify compounds. Figure 2-2 illustrates the absorption spectra of two biologically important molecules, chlorophyll a and oxidized cytochrome c. Chlorophyll a absorbs primarily in the red and blue regions of the spectrum and is clearly distinguishable from the cytochrome c.

Another extremely useful application of this technique is the identification of the light-absorbing molecule (chromophore) associated with a specific light-mediated process, such as photosynthesis. Figure 2-3 compares the absorption spectra for chlorophylls a and b with the action spectrum for photosynthesis. The close correspondence of the absorption spectra with the wavelengths of light active in driving photosynthesis is a clear indication that chlorophyll is the major light-harvesting molecule in photosynthesis.

Fig. 2-2. The absorption spectra of cytochrome c (———) and chlorophyll a (– – –).

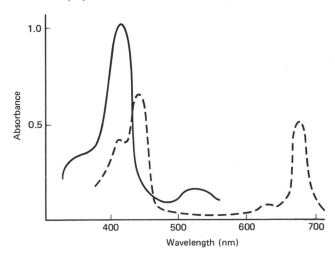

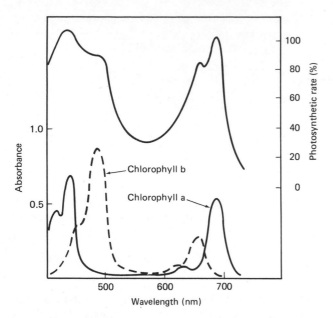

Fig. 2-3. A comparison of the action spectrum for photosynthesis (uppermost plot) with the absorption spectra of chlorophylls a and b (redrawn from Lehninger, 1975).

A further indication of the sensitivity of this technique in distinguishing compounds can be gained by examining the quite dissimilar absorption spectra of chlorophyll a and chlorophyll b (Figure 2-3), two compounds that are almost identical structurally.

The colorimetry exercises described below provide an opportunity to determine the absorbance spectrum for riboflavin, use the same solution to construct a standard curve, and verify the combined Lambert-Beer law. The exercises should occupy no more than one laboratory period.

Colorimetry I. The Absorption Spectrum of Riboflavin

Introduction
Riboflavin (vitamin B_2) is a yellow substance which absorbs light in the visible region. The object of this exercise is to determine its absorption spectrum through colorimetry.

Procedure
1. Obtain from your instructor about 40 ml of a solution of riboflavin having a concentration of 5.31×10^{-5} M.
2. Using a colorimeter tube containing distilled water (called the blank tube), adjust the colorimeter so that the absorbance (A) is zero at 400 nm. In a similar fashion, adjust the instrument so that the absorbance of an opaque substance is infinity at 400 nm. This procedure for "blanking the instrument" varies with manufacturer, and you should consult the laboratory instructor for specific details regarding your instrument.
3. Fill a colorimeter tube nearly to the top with the riboflavin solution, insert it in the sample compartment of the instrument, and determine the absorbance of the solution at 400 nm. Record the reading in Table 2-1.
4. Adjust the wavelength setting of the instrument to 340 nm, blank the instrument with distilled water, and determine the absorbance of riboflavin at this shorter wavelength. Repeat this operation at 10-nm intervals, keeping a record of all readings from 340 to 500 nm (Table 2-1).

Table 2-1.
Absorbance readings for riboflavin

Wavelength (nm)	Absorbance	Wavelength (nm)	Absorbance
340		430	
350		440	
360		450	
370		460	
380		470	
390		480	
400		490	
410		500	
420			

Calculations and Questions

1. Use graph paper to plot a carefully labeled absorption spectrum for riboflavin. Convention calls for plotting the wavelength (in nm) along the abscissa (*x* axis) and the absorbance along the ordinate (*y* axis). This absorption spectrum is a unique characteristic of riboflavin and can be used, along with other criteria, for its identification.

   a. Determine the number of absorption peaks for riboflavin and the absorption maximum for each peak.

   b. Does your absorption spectrum compare favorably with those found in the literature (Clark and Switzer, 1977)? If it does not, speculate as to why it does not.

   c. Look up the absorption spectra for other biological pigments, such as chlorophyll, carotene, or hemoglobin (Armstrong, 1983), and compare them with your absorption spectrum for riboflavin. How do they differ?

2. The Lambert-Beer expression ($A = Ec\ell$) can now be used to calculate the molar extinction coefficient for riboflavin.

   a. Although the relationship holds for all wavelengths, it is best to do the calculation using the absorbance at 450 nm because most literature values for *E* have been determined at that wavelength. The outer diameter of the colorimeter tube should be used as the optical path length (ℓ).

   b. Compare your value with those found in Dawson et al. (1969). Be certain that the units are correct. What might account for variations between your determinations and the literature values?

Colorimetry II. A Standard Curve for Riboflavin

Introduction

One of the most practical applications of the Lambert-Beer law is the use of the standard curve for determining the concentration of solutions. The procedure calls for the preparation of several solutions of riboflavin — the concentration of which are precisely known — and the determination of the absorbance for each solution at a specified wavelength. The resulting data are plotted as a "standard curve" and should show that the absorbance increases as a linear function of the concentration. The concentration of unknown riboflavin solutions can then be determined by comparing absorbance values with those on the standard curve.

Procedure

1. Using the solution of riboflavin from the previous exercise, prepare the six solutions specified in Table 2-2.

2. Calculate the molar concentration for each solution and enter the value in Table 2-2.

3. Set the colorimeter at 450 nm, adjust the absorbance to zero using the contents of tube 1 as a blank, and determine the absorbance for each of the remaining solutions. Carefully record your data in Table 2-2.

4. Obtain a riboflavin solution of unknown concentration from the laboratory instructor, record its code letter, and measure the absorbance at 450 nm.

Calculations and Questions

1. Prepare the standard curve by plotting absorbance at 450 nm (on the ordinate) versus concentration. If the standard curve is linear, it can be used to determine the concentration of other riboflavin solutions. Under normal laboratory conditions it would be necessary to make triplicate determinations of the riboflavin solutions of known concentration and plot the average absorbance for each concentration to assure linearity of the standard curve.

2. Use the standard curve to determine the molarity of your unknown riboflavin solution.

3. Using the data gathered for the standard curve, demonstrate that the molar extinction coefficient for riboflavin is in fact a constant, independent of concentration.

Additional Reading

F. B. Armstrong, *Biochemistry,* 2nd ed., Chap. 18. Oxford University Press, New York, 1983.

J. M. Clark and R. L. Switzer, *Experimental Biochemistry,* 2nd ed. Freeman, San Francisco, 1977.

Table 2-2.
Data table for riboflavin standard curve

Tube	$5.31 \times 10^{-5}\ M$ Riboflavin (ml)	Distilled water (ml)	Concentration (M)	Absorbance at 450 nm
1	0	10.0		
2	2.0	8.0		
3	4.0	6.0		
4	6.0	4.0		
5	8.0	2.0		
6	10.0	0		

R.M.C. Dawson, D. C. Elliot, W. H. Elliot, and K. M. Jones, *Data for Biochemical Research.* Oxford University Press, New York, 1969.

E. B. Kearney, "Flavokinase," in *Methods in Enzymology,* S. P. Colowick and N. O. Kaplan, eds., Vol. II, pp. 640–645. Academic Press, New York, 1955.

A. L. Lehninger, *Biochemistry,* 2nd ed. Worth, New York, 1975.

K. C. Smith, *The Science of Photobiology.* Plenum, New York, 1977.

D. C. Wharton, and R. E. McCarty, *Experiments and Methods in Biochemistry.* Macmillan, New York, 1972.

3 PROTEINS: THE IDENTIFICATION OF A DIPEPTIDE

Although the cell is composed almost entirely of water, it also contains a number of other chemical species which are essential to normal cellular activity. These organic compounds, which are unique to living systems and play a wide variety of dynamic and structural roles in the cell include the proteins, lipids, carbohydrates, and nucleic acids. In the course of this book you will have an opportunity to learn about each of these. The proteins, which are considered here and in Chapters 4 and 5, are among the most important biomolecules both in terms of their abundance (15 per cent by weight) and in terms of their varied cellular functions. It has been estimated that a living cell contains at least 3,000 different kinds of proteins acting in a variety of roles—for example, they may serve as biochemical catalysts, energy sources, molecular messengers, and structural components, to name just a few.

Amino Acids

The fundamental building blocks of protein are called amino acids. These relatively small molecules are characterized by the presence of an α-amino group, an α-carboxyl group, and a unique R group (Figure 3-1). There are 20 different amino acids commonly found in living tissue. All of the amino acids are structurally identical except for the R groups, which differ in polarity, size, solubility, and electrical charge.

Protein Structure

Native protein, as it occurs in the cell, is a high-molecular-weight polymer of amino acids. The individual residues are linked in a linear sequence to form an unbranched polymer ranging from 50 to several thousand amino acids in length; thus the molecular weight of protein ranges from about 6,000 to several million daltons. The linkage between two adjacent amino acids is called a peptide bond and is formed by the condensation of the α-carboxyl group of one amino acid with the α-amino group of another. Figure 3-2 illustrates how such a bond is formed between isoleucine and phenylalanine.

The molecule on the right side of the reaction in Figure 3-2 is called a dipeptide because it contains two amino acids. Other simple peptides are similarly referred to as tripeptides, tetrapeptides, and so forth. Notice that the dipeptide has a free amino group and a free carboxyl group, a property it shares with all amino acids and other peptides. The free amino group of a peptide is referred to as the N terminus and is written to the left, and the free carboxyl group is called the C terminus. The hypothetical pentapeptide shown in Figure 3-3 illustrates these features and provides a more accurate picture of the bond angles between atoms of the peptide.

A typical protein is much longer than the peptide illustrated in Figure 3-3 and would most likely not be in the linear, extended form implied by the diagram. The interaction of the R groups with one another and with the aqueous medium causes the formation of innumerable kinks, twists, and bends in the protein molecule, giving it a unique three-dimensional conformation. Since it has been shown that the biological activity of a protein depends upon its three-dimensional conformation and that the conformation depends upon the

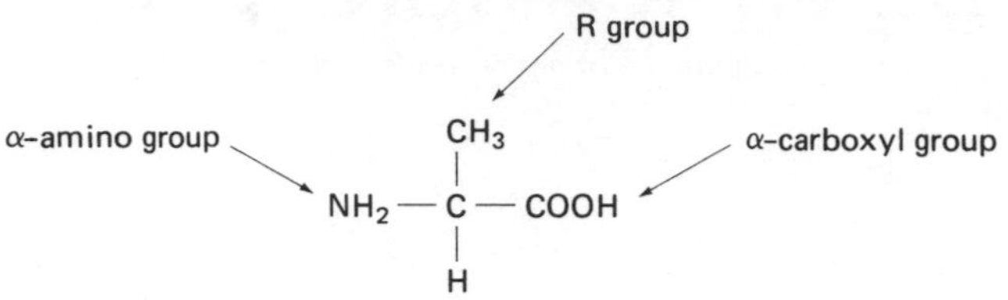

Fig. 3-1. The amino acid alanine, for which R is a methyl group.

amino acid sequence, it should be no surprise that the sequence and composition of proteins are of particular interest to biochemists.

Determination of Protein Composition

A procedure often used to determine protein composition and sequence involves partial or total hydrolysis of the protein followed by a detailed analysis of the reaction products. Hydrolysis can be achieved by treatment with acid or base at high temperature or with proteolytic enzymes at room temperature. Separation and identification of the peptides and free amino acids are

usually accomplished by chromatography, electrophoresis, or a combination of the two.

The exercise described below is a simplified version of this procedure. You will be provided with a dipeptide of unknown composition and expected to determine its identity. Your job is to hydrolyze the dipeptide in acid and then identify the amino acids by paper chromatography. It should take no more than two laboratory periods if you divide up your work in the following way:

Day 1. Start acid hydrolysis of dipeptide.
Prepare and spot chromatograms with amino acid standards.
Day 2. Spot papers with hydrolyzed dipeptides.
Proceed with chromatography.

Dipeptide Identification.
Acid Hydrolysis and Paper Chromatography

Introduction

The procedure calls for overnight acid hydrolysis of a small amount of the dipeptide at 100°C. The combination of low pH and high temperature hydrolyses the

Fig. 3-2. The formation of a peptide bond between two amino acids.

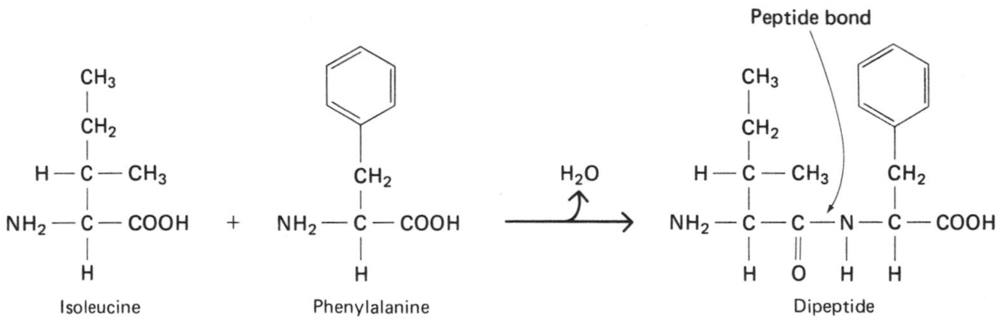

Fig. 3-3. A pentapeptide with unspecified R groups. This structure illustrates the juxtaposition of substituent groups and provides a good picture of the actual bond angles.

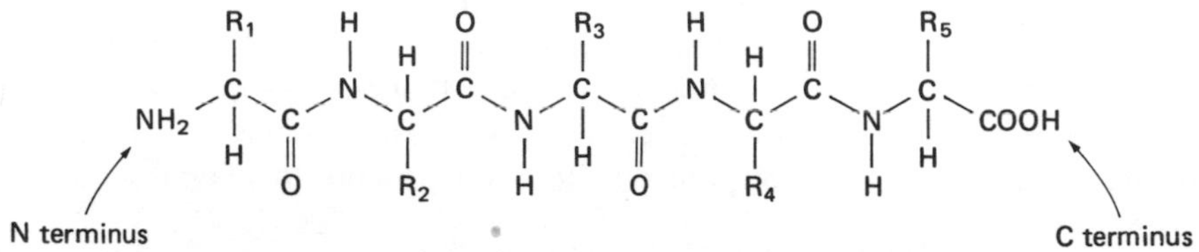

peptide bonds and releases the amino acids into solution. A side effect of these conditions is the partial or total destruction of some amino acids, most notably tryptophan, serine, and threonine.

The acid-hydrolysed dipeptide is applied to chromatography paper, which is then placed in a large glass jar containing a small amount of an appropriate solvent mixture. The solvent is absorbed by the paper and moves upward by capillary action, dissolving the amino acids and carrying them upward. Separation is achieved as a result of the differential solubility of the amino acids in the solvent: those that are very soluble ascend more rapidly than do those that are less soluble. When the solvent migration is terminated, the final position of the amino acids is determined with a locating solution and the identity of unknowns determined by comparison with the position of known amino acid standards.

Procedure

1. Total hydrolysis of the dipeptide
   a. Obtain from your laboratory instructor a bottle containing the unknown dipeptide identified by a code letter. Record the letter and weigh out 10 mg of the material. You will need to use an analytical balance for this procedure and should consult your laboratory instructor for specific instructions.
   b. Place the dipeptide on a clean watch glass and dissolve it in 0.3 ml of 6.0 *M* HCl.
   c. Transfer all of this acidic solution into several sealed capillary tubes in the following manner. Seal the end of a capillary tube (1.6 to 1.8 $\times$ 100 mm) with a Bunsen burner. Dip the open end of the tube into the dipeptide solution. As the glass cools, the air pressure inside the tube drops and the solution is drawn into the tube. When this operation is completed, tip the tube up and tap it gently to move the liquid away from the open end. Seal the open end of the capillary with the Bunsen burner, and check to be sure that there are no leaks.
   d. Label the tubes and incubate them in an oven at 100°C for 12 to 16 hours.
2. Preparation of the chromatograms

   As indicated in Table 3-1, it is necessary to prepare two identical sets of chromatograms and run them in two different solvent systems. This dupli-

Table 3-1.

Suggested combinations of amino acids and chromatography solvents

1. Amino acid series A : Solvent system I
2. Amino acid series B : Solvent system I
3. Amino acid series A : Solvent system II
4. Amino acid series B : Solvent system II

cation of effort is justified because some amino acids separate better in one solvent system than another. The resulting data from both systems provide sufficient information to make your final determinations.

   a. Prepare four sheets of Whatman No. 1 chromatography paper (21 $\times$ 26 cm). Whenever handling the paper, take precautions to keep it as clean as possible because fingerprints contain amino acids. This can best be done by wearing disposable plastic gloves and placing the sheets on or between pieces of clean paper toweling.
   b. Designate the bottom of each chromatogram by drawing a single *pencil* line (no ink) parallel to the long axis of the paper, 3 cm from the edge. Place a small dot on the line every 2 cm across the sheet and number the dots 1 through 12. The dots indicate where the amino acids will be applied and are referred to as the origin of the chromatogram.
   c. In the lower left hand corner, identify each of the sheets with your name, today's date, and the four different amino acid:solvent system combinations specified in Table 3-1 (Figure 3-4).
3. Application of samples to chromatograms
   a. Amino acid standards
      1) Using a micropipette or calibrated capillary,* apply 20 μl of each amino acid to the numbered positions on the chromatograms according to the protocol shown in Table 3-2. Do not add anything to positions 4 and 9 at this time, as these are reserved for the hydrolysed dipeptide.
      2) As you apply the amino acid solutions to the paper, endeavor to keep the spots as small

*A capillary tube (1.6 to 1.8 $\times$ 100 mm) can be calibrated to deliver about 20 μl by placing a mark 2.0 cm from the end of the tube.

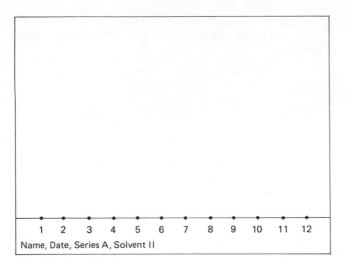

Name, Date, Series A, Solvent II

Fig. 3-4. A chromatogram set up to run the series A amino acids in solvent II.

Table 3-2.
Protocol for applying amino acids to chromatograms

Amino acid series A		Amino acid series B	
Position	Amino acid	Position	Amino acid
1	Isoleucine	1	Leucine
2	Lysine	2	Arginine
3	Valine	3	Proline
4	Dipeptide	4	Dipeptide
5	Glutamic acid	5	Asparagine
6	Histidine	6	Serine
7	Alanine	7	Methionine
8	Glutamine	8	Aspartic acid
9	Dipeptide	9	Dipeptide
10	Tyrosine	10	Threonine
11	Cysteine	11	Phenylalanine
12	Trytophan	12	Glycine

as possible (2 to 3 mm in diameter) by adding a small amount at a time. This can best be done by gently pressing the tip of the pipette onto the paper and removing it as soon as some liquid has flowed into the paper. After each application, let the spot dry before adding more liquid. Drying can be hastened by the use of a hair dryer or heat gun. If you have not spotted a chromatogram before, it would be wise to practice with water and a scrap of Whatman No. 1 paper.

NOTE: Be sure to use a clean pipette or capillary for each amino acid.

   b. Hydrolysed dipeptide
     1) After the sealed capillary tubes are removed from the oven, permit them to cool to room temperature.
     2) Open the tubes by making a small scratch in the glass near one end with a triangular file and then snap the tube between your fingers.
     3) Apply a single spot of the hydrolysed dipeptide at positions 4 and 9 of each chromatogram. It may be necessary to open both ends of the tubes to do this.
   c. After all the samples have been applied, store the chromatograms between pieces of clean paper toweling.
4. Preparation of chromatography solvents
   a. Solvent I, *n*-butanol : formic acid : water (100 : 30 : 25)
     1) Combine 200 ml of *n*-butanol with 60 ml of formic acid and 50 ml of distilled water. This solvent tends to deteriorate with time and should be used immediately after it has been prepared.
     2) Pour the solvent into two chromatography jars to a depth of about 1 cm. Label each of

Fig. 3-5. Chromatography set up for solvent II just before the chromatogram is added.

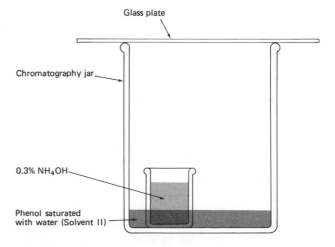

Glass plate

Chromatography jar

0.3% NH$_4$OH

Phenol saturated with water (Solvent II)

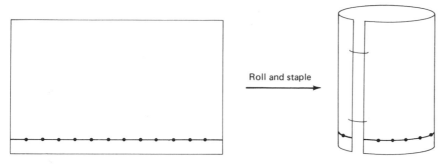

Roll and staple

Fig. 3-6. Final preparation of the chromatogram prior to chromatography.

the jars and cover it with a glass plate or watch glass.

b. Solvent II, phenol saturated with water in an ammonia atmosphere.

CAUTION: Phenol can cause serious burns and should be handled with caution. If you get some on your skin, rinse it off immediately with ethyl alcohol and notify the laboratory instructor.

1) Pour the aqueous phenol solution into two chromatography jars to a depth of about 1 cm. Place a small beaker containing 20 ml of 0.3% (v:v) NH_4OH on the bottom of the jar (Figure 3-5).

2) Label each of the jars and cover with a glass plate or watch glass.

5. Chromatography

a. Roll and staple the chromatography sheets in the form of a cylinder, being certain that the edges do not overlap (Figure 3-6). The origin of the chromatogram should be at the end of the cylinder.

b. Place one chromatogram in each jar (sample spots down; Figure 3-7) in accordance with the protocol described in Table 3-1. Replace the jar lids and let the solvent run until it is about 1 to 2 cm from the top of the paper (3 to 5 hours).

NOTE: If possible these procedures should be carried out in a fume hood to minimize the inhalation of solvent vapors.

c. Remove the papers from the jars, draw a pencil line along the solvent front, and hang the chromatograms in the hood to dry overnight.

6. Development of the chromatogram

a. Spray each of the sheets with a ninhydrin solution and then place in an oven at 100°C for five minutes. The amino acids should appear as blue or purple spots with the exception of proline, which will be yellow (Figure 3-8).

Fig. 3-7. Rolled chromatogram in the chromatography jar. The glass lid assures a saturated atmosphere inside the jar.

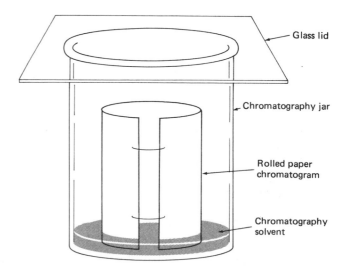

Glass lid

Chromatography jar

Rolled paper chromatogram

Chromatography solvent

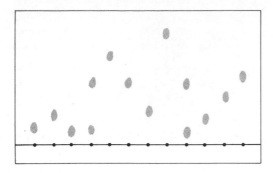

Fig. 3-8. A chromatogram after treatment with the locating solution.

b. Circle each of the spots on the chromatogram and measure the distance it has moved from the origin. When you make these measurements, choose some reference point on the amino acid spot—for example, the leading edge or the geometric center—and use that consistently throughout. Also determine the distance moved by the solvent front above each amino acid. These data should be recorded in your notebook in a series of tables similar to the one illustrated in Figure 3-9.

c. Prepare an accurately labeled sketch of each of the chromatograms for your notebook. The diagram should indicate exact location of the origin, solvent front, and all spots, including any extraneous ones.

Calculations and Questions

1. Determine R_f values for all amino acids on the chromatograms using the following equation:

$$R_f = \frac{\text{distance from origin to spot}}{\text{distance from origin to solvent front}}$$

Enter your results in the chromatography data sheets (Figure 3–9).

2. Examine your data and answer the following questions:
   a. Are the R_f values for a single amino acid identical in both solvent systems? How much variation is there?
   b. Are the R_f values for the amino acids of the dipeptide consistent within a single solvent system? How much variance is there?

c. Consult the Additional Reading regarding the properties of amino acids and their behavior in partition chromatography. Have they separated as you would expect them to? Explain your answer.

3. Compare the R_f values of the standard amino acids with those of the hydrolysed dipeptide and determine the identity of the two unknown amino acids (positions 4 and 9). Use your data from both solvent systems as well as the color and shape of the spots in your determination. Report your conclusions and the reasoning behind them.

4. If you were to find a single amino acid spot at positions 4 and 9 of all four chromatograms, how might you explain the result?

5. What do the results of this exercise tell you about the sequence of the dipeptide?

6. Investigate some of the literature cited in Additional Reading to determine why it is necessary to run

Fig. 3-9. Suggested format for organizing the amino acid chromatography data in your notebook.

CHROMATOGRAPHY DATA SHEET	NAME:		
	DATE:		
SOLVENT SYSTEM-n-butanol: formic acid: water (100:30:25)			
AMINO ACID SERIES B			
DISTANCE (cm)			
SAMPLE	SPOT	SOLVENT FRONT	Rf VALUE
1. leu	13.5	16.9	0.90
2. arg	3.2	16.9	0.19
3. pro			
4. a.			
b.			
5. asn			
6. ser			
7. met			
8. asp			
9. a.			
b.			
10. thr			
11. phe			
12. gly			

amino acids standards each time you attempt to identify an unknown by partition chromatography.

Additional Reading

F. B. Armstrong, *Biochemistry,* 2nd ed. Chaps. 6 and 7. Oxford University Press, New York, 1983.

J. M. Clark Jr., *Experimental Biochemistry.* Freeman, San Francisco, 1964.

J. M. Clark Jr. and R. W. Switzer, *Experimental Biochemistry, 2nd ed. Freeman, San Francisco, 1977.*

A. L. Lehninger, *Biochemistry,* 2nd ed. Worth, New York, 1975.

D. T. Plummer, *An Introduction to Practical Biochemistry,* McGraw-Hill, New York, 1971.

G. Zweig and J. Sherma, eds, *Handbook of Chromatography.* Vols. I and II. CRC Press, Cleveland, 1972.

4 PROTEINS: THE EXTRACTION AND PURIFICATION OF WHEAT GERM ACID PHOSPHATASE

Protein Isolation Procedures

Proteins play such a variety of important roles in the living cell that it should be no surprise that an enormous amount of time and effort has been spent in elucidating the structure and function of these versatile macromolecules. An absolute prerequisite to meaningful progress in the study of any protein is an ample supply of highly purified material. Unfortunately, few if any proteins exist in nature at the desired concentration or at the degree of purity required for such detailed analysis. Consequently it has been necessary that laboratory procedures be developed for the extraction, concentration, and purification of cellular proteins.

The development of these procedures, which has taken a considerable period of time, has been hampered by the fact that the concentration of any given protein in a cell is usually extremely low (less that 0.01 per cent of the total cell mass) and further complicated by the presence of many other macromolecules (nucleic acids, carbohydrates, etc.), which must be eliminated. The task of a biochemist who chooses to develop an isolation technique for large amounts of purified protein is a difficult one and would typically include the following considerations:

1. The development or selection of a simple assay procedure that specifically demonstrates the presence and concentration of the protein in question.
2. The choice of material from which the protein can be isolated. The overriding factors in this choice typically are the absolute concentration of the protein in a given natural source and the cost and availability of the material.
3. The selection of a cell-disruption technique appropriate to the cell or tissue type being used. Some materials, such as plant cells, for example, require harsher techniques than others and run the risk of physical disruption of the protein.
4. The development of extraction conditions that maintain a pH, temperature, and ionic environment conducive to maintaining the native configuration of the protein. Extremes of these conditions increase the likelihood of inactivating the protein by denaturation.
5. The selection of a series of precipitation and centrifugation procedures that maximize the concentration of the chosen protein and minimize that of the other contaminating molecules.

Biochemists have, through a process of trial and error, addressed these considerations and have developed an impressive array of techniques for the extraction and purification of proteins. A point that should be emphasized is that there is no single, foolproof protein isolation technique uniformly applicable to all situations. A biochemist with the task of isolating a protein that has not been purified previously develops an isolation scheme by first trying several of the most reliable techniques for that general class of proteins. After each step in the scheme, the effectiveness of the technique is evaluated and subsequent isolation schemes modified accordingly. Thus, any given isolation procedure has in all probability arisen through the application of well-established techniques tested by trial and error.

Introduction

The Enzyme

The protein to be extracted and purified in the following exercise is an enzyme called wheat germ acid phosphatase. This same enzyme is the subject of detailed analysis in Chapter 5. The systematic name of acid phosphatase is orthophosphoric-monoester phosphohydrolase (acid optimum), E.C. 3.1.3.2. The enzyme catalyzes the release of inorganic phosphate (P_i) from such diverse substances as ATP, glucose-6-phosphate, and glycerol-2,3-diphosphate at pH values ranging from 4 to 6 (Figure 4-1).

Acid phosphatase is found widely distributed among plants, animals, and microorganisms, although its precise biochemical role remains unclear. Wheat germ has been selected as a source for the enzyme because it is inexpensive and readily available and contains high levels of the enzyme.

Extraction Procedures

The procedures detailed below for the extraction of wheat germ acid phosphatase are representative of a typical protein isolation scheme. The initial water extraction of the enzyme (along with a number of other water-soluble proteins, carbohydrates, etc.) is accomplished by soaking the wheat germ in cold distilled water. The insoluble material is removed by a brief centrifugation, which yields a supernatant containing the enzyme and a variety of other soluble contaminants. These contaminants (protein and otherwise) are in turn removed by a series of selective precipitation and centrifugation steps to separate the soluble (supernatant) from the insoluble (pellet) fraction.

Salt fractionation, or salting out, of proteins is often used at the beginning of a protein isolation scheme because it is a convenient method for separating large groups of proteins in one or two steps. The solubility of proteins in aqueous solution depends on the number of hydrophilic amino acids in the protein and on the ionic strength of the solution. When the ionic strength is gradually increased by the addition of manganese chloride or ammonium sulfate, fewer and fewer water molecules are available to solubilize the proteins and eventually the proteins begin to come out of solution (salt out). Thus, by carefully controlling the salt concentration, one can selectively precipitate whole groups of proteins while leaving others in solution. Previous experience has demonstrated that acid phosphatase is soluble in a 35% ammonium sulfate solution and insoluble in 57% ammonium sulfate.

The subsequent heat treatment and methanol fractionation operate in much the same way, precipitating some proteins and leaving others in solution. Since this

Fig. 4-1. A typical acid phosphatase reaction, in which glucose-6-phosphate is degraded to glucose and inorganic phosphate.

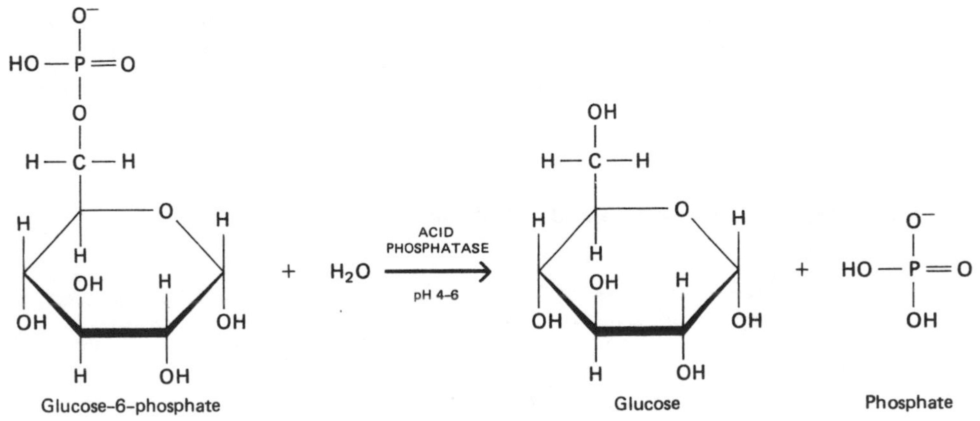

procedure, like all others, is not foolproof, it is highly advisable to verify the presence of the enzyme in each fraction before moving on to subsequent steps.

The completion of this purification scheme is accomplished by dialysis, a technique designed to remove small-molecular-weight contaminants. The enzyme extract is enclosed within a dialysis sac (a semipermeable membrane) and suspended in a large volume of dilute salt. As the solution is gently stirred, the small molecules pass out through the pores, leaving the large-molecular-weight enzyme inside the sac. At equilibrium, the majority of the small molecules are removed in this fashion.

Assay Procedures

The method of choice for protein measurement during enzyme purification is the Biuret reaction. The Biuret reagent contains copper ions in alkaline solution, and these ions form colored complexes with the peptide linkages of protein. The concentration of protein is then determined colorimetrically. This method is advantageous because it is quick and easy and because the high concentrations of ammonium sulfate typical of early fractionation steps do not interfere with the color reaction.

The assay for acid phosphatase activity depends upon tbe fact that the enzyme catalyzes the hydrolytic release of *para*-nitrophenol and inorganic phosphate from the artificial substrate *p*-nitrophenyl phosphate (Figure 4-2). Under alkaline conditions one of the reaction products (*p*-nitrophenol) absorbs light strongly at 405 nm, and its concentration can be determined colorimetrically. The enzyme assay calls for combining the enzyme and substrate for a brief period of time, stopping the reaction, and determining the amount of substrate produced per unit time.

Procedure

The overall organization of this section is based upon what a group of students can be expected to accomplish in three successive 4-hour laboratory sessions. This assumes that you have read all the preliminary material before coming to the laboratory and that all of the necessary solutions and equipment are readily available. A brief synopsis of what you should be able to complete in each day follows:

Day 1. Extraction of acid phosphatase through supernant V (SV). This fraction is stable and may be stored frozen for several weeks with little or no loss in enzyme activity.

Day 2. Protein and phosphatase assays on all samples (SI, SII, etc.) to assess the effectiveness of the enzyme isolation procedure.

Day 3. Final purification of the enzyme starting with SV and continuing through the dialysis step.

Unless otherwise indicated, all procedures should be carried out at 0 to 4°C to minimize heat denaturation and protease damage to the enzyme. The flow sheet shown in Figure 4-3 illustrates in an abbreviated fashion all the steps in the procedure and may be used as a checklist throughout the process.

First Day

1. Water extraction of the enzyme from wheat germ
   a. Transfer 50 g of wheat germ to a 250-ml beaker surrounded by crushed ice in an ice bucket.
   b. Add 200 ml of cold distilled water, mix, and allow to stand for 30 minutes with occasional stirring. This procedure will dissolve the water-soluble proteins, including the acid phosphatase.

Fig. 4-2. The hydrolysis of *p*-nitrophenyl phosphate by the enzyme acid phosphatase.

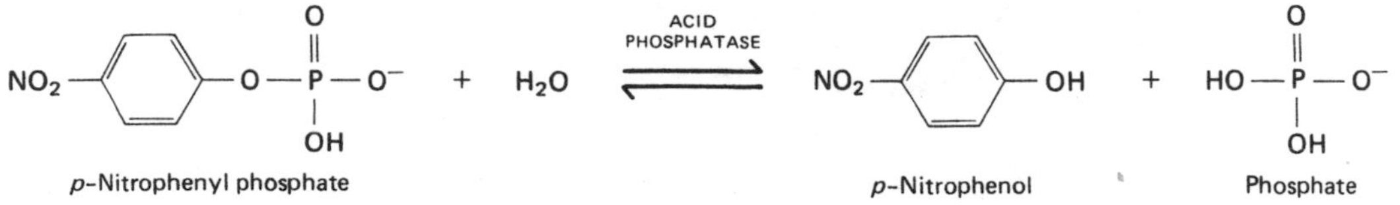

p-Nitrophenyl phosphate   *p*-Nitrophenol   Phosphate

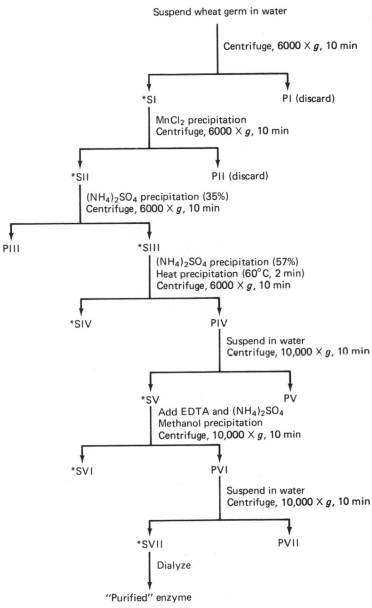

Suspend wheat germ in water

Centrifuge, 6000 × g, 10 min

*SI   PI (discard)

MnCl$_2$ precipitation
Centrifuge, 6000 × g, 10 min

*SII   PII (discard)

(NH$_4$)$_2$SO$_4$ precipitation (35%)
Centrifuge, 6000 × g, 10 min

PIII   *SIII

(NH$_4$)$_2$SO$_4$ precipitation (57%)
Heat precipitation (60°C, 2 min)
Centrifuge, 6000 × g, 10 min

*SIV   PIV

Suspend in water
Centrifuge, 10,000 × g, 10 min

*SV   PV

Add EDTA and (NH$_4$)$_2$SO$_4$
Methanol precipitation
Centrifuge, 10,000 × g, 10 min

*SVI   PVI

Suspend in water
Centrifuge, 10,000 × g, 10 min

*SVII   PVII

Dialyze

"Purified" enzyme

Fig. 4-3. A flow sheet illustrating the entire isolation procedure for wheat germ acid phosphatase. Asterisks indicate fractions for which a small amount should be set aside for later analysis, P indicates a pellet, and S indicates a supernatant.

c. Centrifuge the mixture at 6000 × g for 10 minutes to separate the heavier cellular material from the soluble proteins.

NOTE: Operation of the centrifuge according to proper guidelines is absolutely essential for safety in the laboratory and for reliable results.

1) Be sure that the centrifuge tubes and inserts you are using are those designed for your particular centrifuge and rotor.
2) Do not fill the centrifuge tubes to more than two thirds of their total capacity.
3) Be certain that, as you place filled tubes in the rotor, there is a balancing tube of equal weight on the opposite side of the rotor.
4) Consult your laboratory instructor for specific instructions regarding the operation of the centrifuge.

d. When the centrifugation is complete, remove the tubes and carefully decant the supernatant (SI) into a chilled graduated cylinder, recording its total volume. The pellet (PI) at the bottom of the tube may be discarded.

e. The water extract (SI) contains the enzyme as well as a number of other soluble molecules. Set aside a small volume (1.0 ml) of SI for subsequent protein and enzyme analysis.

2. Salt fractionation of the enzyme with manganese chloride
   a. Transfer SI to a large beaker in an ice bath on top of a stirring motor. Place a magnetic stirring bar in the solution and turn the motor on low speed to stir the mixture gently (Figure 4-4).
   b. Add 1.0 M $MnCl_2$ (2.0 ml for every 100 ml of SI) to the mixture. The $MnCl_2$ solution should be added slowly and with gentle mixing.
   c. Centrifuge at 6000 × g for 10 minutes, reserving the enzyme-containing supernatant (SII) and discarding the pellet (PII).
   d. Determine the total volume of SII and set aside a small amount (1.0 ml) for later analysis.

3. Salt fractionation of the enzyme with ammonium sulfate (35% cut)
   a. Transfer SII to a 500-ml beaker situated in a gently stirring ice bath (Figure 4-4) and add 54 ml of cold saturated $(NH_4)_2SO_4$ (pH 5.5) for every 100 ml of SII.

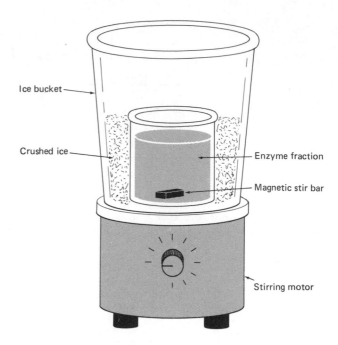

Ice bucket

Crushed ice

Enzyme fraction

Magnetic stir bar

Stirring motor

Fig. 4-4. A suggested setup for continually stirring the solution at ice-bath temperatures.

b. The addition of ammonium sulfate should be done slowly to avoid localized denaturation of protein. This can be accomplished best by using a 10-ml pipette to add the salt solution to the slowly stirring supernatant over a 10 to 15 minute period. If there is any sign of foaming, the rate of stirring should be reduced because foaming denatures protein and may render the enzyme inactive.

c. Centrifuge at 6000 × g for 10 minutes, saving the supernatant (SIII) and discarding the pellet (PIII).

NOTE: Ammonium sulfate is extremely corrosive to aluminum. If it is spilled on the centrifuge rotor, be certain to rinse it thoroughly with cold tap water.

d. Measure the volume of SIII, which is now 35% with respect to ammonium sulfate and contains the bulk of the acid phosphatase. Set aside a small sample (1.0 ml) of SIII for subsequent analysis.

4. Precipitation of the enzyme by ammonium sulfate (57% cut) and heat treatment (60°C)

a. For every 100 ml of SIII, add 79 ml of cold saturated ammonium sulfate (pH 5.5) in precisely the same fashion as described above.

b. When all of the ammonium sulfate has been added, transfer the beaker to a hot water bath (65 to 70°C) and swirl the mixture gently until the temperature reaches 60°C. Maintain the mixture at that temperature for exactly 2 minutes and then immediately return the beaker to the ice bath, cooling the mixture to 5°C as rapidly as possible. The temperature of the mixture should be monitored throughout this procedure with a thermometer, which can also be used (with care) to mix the solution.

c. Centrifuge at 6000 $\times$ g for 10 minutes to collect the enzyme, which should now be in the insoluble portion (PIV).

d. Retain the pellet (PIV) but do not discard the supernatant (SIV) until the presence of acid phosphatase in PIV has been verified by enzyme analysis. The supernatant should be stored frozen until PIV has been tested for enzyme activity.

5. Water extraction of the enzyme from PIV

a. Add a small volume (5 to 10 ml) of cold distilled water to each centrifuge tube containing an enzyme pellet (PIV). Resuspend the pellets by scraping and agitating the precipitated material with a blunt glass rod.

b. When a uniform suspension has been achieved, measure the total volume and add enough cold distilled water to give a final volume equal to one third the volume of the $MnCl_2$ supernatant (SII).

c. Centrifuge the suspension at 10,000 $\times$ g for 10 minutes, keeping the supernatant (SV) and discarding the pellet (PV).

d. Measure the volume of SV and save a small amount (1.0 ml) for later analysis.

e. Transfer SV to a plastic bottle and store it in the freezer. The acid phosphatase may be stored in this condition for several weeks with little decrease in enzyme activity.

6. Place all of the 1.0-ml samples (SI, SII, etc.) you have collected in the freezer.

Second Day
It is your task at this point to evaluate the effectiveness of the protein isolation procedure through an analysis of all the fractions thus far collected (SI, SII, etc.). Test each sample for the amount of protein present and the level of enzyme activity. These and other relevant data should be entered in the enzyme purification table (Table 4-1), from which the per cent recovery at each step can be determined.

1. Protein concentration — Biuret test

a. Protein standard curve using bovine serum albumin

1) Obtain a sample of bovine serum albumin (BSA) having a concentration of 10 mg/ml and prepare a series of 1.0-ml solutions containing a range of protein concentration from 0 to 10 mg/ml.

2) To each 1.0-ml sample, add 4.0 ml of Biuret reagent, mix well, and allow to stand at room temperature for 30 minutes.

3) Determine the absorbance of each sample at 540 nm (A_{540}).

4) Prepare a standard curve by plotting A_{540} versus protein concentration (mg/ml).

b. Determination of protein concentration in enzyme fractions

1) Set up a series of labeled test tubes containing 0.1-ml samples of the various enzyme fractions you plan to assay (SI, SII, etc.).

2) Add 0.9 ml of distilled water to each sample, bringing it to a final volume of 1.0 ml.

3) Add 4.0 ml of Biuret reagent to each sample and let stand at room temperature for 30 minutes.

4) Determine the absorbance at 540 nm for each sample and, using the protein standard curve, determine the protein level (mg/ml) for each fraction.

5) Enter this information plus the total protein per sample information in the appropriate columns of Table 4-1.

c. Notes

1) If the absorbance is beyond the sensitivity of the standard curve, it will be necessary to repeat the assay at a more appropriate dilution of the sample.

2) As you do the final calculations for protein concentration, be mindful of the dilution factor introduced during the assay. For example, if a 0.1-ml sample has been diluted with

Table 4-1.
Enzyme purification table for wheat germ acid phosphatase

| Fraction | Vol (ml) | Protein | | | Enzyme Activity | | | | |
		Sample (ml)	Protein (mg/ml)	Total protein (mg)	Sample (ml)	Activity (units/ml)	Total Enzyme (units)	Specific Activity (units/mg protein)	Percent recovery
SI									
SII									
SIII									
SIV									
SV									
SVI									
SVII									
Pure enzyme									

0.9 ml of water prior to the addition of Biuret reagent, the sample has been diluted ten times and the protein concentration must be adjusted accordingly.

2. Enzyme activity—acid phosphatase assay

The level of acid phosphatase is determined by an assay which uses the artificial substrate p-nitrophenyl phosphate. To assay a single sample, set up two reaction tubes identical with respect to substrate concentration, temperature, etc., and add enzyme to one tube and an equal volume of water to the other. The reaction is terminated after 5 minutes by adding potassium hydroxide to each tube. The KOH stops the reaction by denaturing the enzyme and converts the p-nitrophenol to its colored form. The contents of the tubes are compared colorimetrically, and the amount of p-nitrophenol produced per minute is used as a measure of enzyme activity.

a. Acid phosphatase assay

1) Prepare two labeled test tubes (A and B) each containing the following reaction mixture:

0.5 ml of 1.0 M sodium acetate buffer (pH 5.7)
0.5 ml of 0.1 M MgCl$_2$
0.5 ml of 0.05 M p-nitrophenyl phosphate
3.3 ml of distilled water

2) Mix the contents of each tube and place them in a 37°C water bath. Allow the temperature to equilibrate for 5 minutes.

3) Add 0.2 ml of distilled water to tube A (blank), mix the contents, and return the tube to the water bath. Add 0.2 ml of the enzyme sample to tube B, start a stopwatch, mix, and return the tube to the water bath.

4) When the reaction has proceeded for 5 minutes, immediately add 2.5 ml of 0.5 M KOH to each reaction tube.

5) If the contents of the tubes appear cloudy due to the presence of precipitated protein, centrifuge the samples for 10 minutes at top speed in a desktop clinical centrifuge. Discard the precipitate.

6) Using the contents of tube A as a blank, determine the absorbance of tube B at 405

nm. If the absorbance is greater than 1.0, the assay must be repeated using a more dilute concentration of the enzyme because such a reading is beyond the sensitivity of the colorimeter.

7) Convert the absorbance readings at 405 nm to micromoles of p-nitrophenol formed per minute, using the extinction coefficient for p-nitrophenol (18.8×10^3 liter mole^{-1} cm^{-1}), which assumes a path length of 1.0 cm.

b. Determine the acid phosphatase activity (units/ml) and total phosphatase activity (units) for all samples (SI, SII, etc.) and enter the data in Table 4-1. One unit of acid phosphatase is defined as the amount of enzyme that catalyzes the hydrolysis of 1.0 micromole of p-nitrophenyl phosphate per minute at 37°C.

c. Notes:

1) The method described above is for a single enzyme sample and should be expanded to test all the fractions (SI, SII, etc.) for phosphatase activity. With proper organization, you should be able to set up a series of reaction tubes, start them in sequence (1 minute apart), and stop them in the same sequence 5 minutes later. A single blank tube is sufficient.

2) As indicated in the protein determination notes, you should keep track of sample dilutions in order to accurately calculate enzyme activity in the samples.

Third Day

Remove SV from the freezer and thaw it rapidly in lukewarm water. Proceed with the remaining steps in the extraction procedure.

1. Methanol fractionation of the enzyme

a. Consult Table 4-1 to determine the protein concentration of SV, and if it is higher than 5 mg/ml, adjust it down to that level with distilled water. Measure and record the final volume.

b. For every 1.0 ml of SV, add 0.11 ml of 0.2 M disodium ethylenediaminetetraacetic acid (EDTA).

c. For every 1.0 ml of SV, add 0.05 ml of saturated ammonium sulfate (pH 5.5). Record the volume of this solution.

d. Transfer the solution to a stirring ice bath and slowly add 1.75 ml of very cold ($-20°C$) methanol for each milliliter of enzyme preparation (SV plus additions just made).

e. Centrifuge at $10,000 \times g$ for 10 minutes, keeping the pellet (PVI) and discarding the supernatant (SVI) after saving 1.0 ml for later analysis.

2. Water extraction of the enzyme from PVI

a. Suspend the pellet (PVI) in 10 ml of cold distilled water with the aid of a blunt glass rod.

b. Centrifuge at $10,000 \times g$ for 10 minutes, decant the supernatant (SVII), and save it on ice.

c. Suspend the pellet (PVII) in 10 ml of cold distilled water and centrifuge at $10,000 \times g$ for 10 minutes. Combine this supernatant with SVII and discard the pellet.

d. Measure and record the volume of the combined supernatants (SVII). Place about 1.0 ml of SVII in the freezer for later analysis.

3. Dialysis of the enzyme

a. Obtain a piece of dialysis tubing approximately 40 cm in length (1.3 cm in diameter) and place it in a beaker of distilled water for 5 to 10 minutes. This treatment hydrates the cellulose tubing and makes it flexible.

b. Rub the tubing between your fingers to separate the side walls and open the lumen. When the dialysis sac is opened fully, tie a double knot in one end.

c. Using a Pasteur pipette, carefully transfer SVII into the dialysis sac, being careful not to puncture the tubing. Tie two knots in the open end, trapping a small amount of air inside the sac.

d. Transfer the filled dialysis sac to a large beaker or flask containing 1000 ml of cold 5.0 m*M* sodium EDTA. Place the container in a coldroom or refrigerator at 4°C and mix it gently overnight with a magnetic stirring motor (Figure 4-5).

e. After removing the sac from the dialysis solution, open one end with clean stainless steel scissors and transfer the contents to a glass test tube.

f. This solution is your purified acid phosphatase

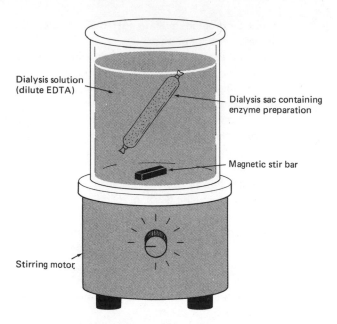

Dialysis solution
(dilute EDTA)

Dialysis sac containing
enzyme preparation

Magnetic stir bar

Stirring motor

Fig. 4-5. A recommended setup for overnight dialysis. This apparatus should be placed in a coldroom or refrigerator.

sample and is used for all subsequent enzyme experiments. Measure the volume of the enzyme sample and store it frozen in a carefully labeled, covered test tube.

4. Use the Biuret and acid phosphatase assays to determine protein and enzyme levels in SVI, SVII, and the purified enzyme sample. Enter these data in Table 4-1.

Calculations and Questions

1. Enter all remaining information (volumes, etc.) in Table 4-1.
   a. Calculate specific activity (units of enzyme per mg of protein) for each fraction.
   b. Assuming that the total number of enzyme units in SI represents 100 per cent recovery, calculate the per cent recovery for each subsequent fraction.
   c. With each subsequent step in the isolation procedure, what happens to the amount of total protein in each fraction? What happens to the specific activity in each fraction? How can you explain these two trends?
   d. At which point in the isolation procedure (35% ammonium sulfate cut, methanol precipitation, etc.) did you realize the greatest per cent recovery of enzyme? If you were planning to improve the isolation procedure to give a higher yield, which step in the procedure would you concentrate on? Explain your answer.

2. Why is it necessary to maintain low temperatures throughout the extraction procedure?

3. The level of acid phosphatase present in the supernatant after the 35% ammonium sulfate cut (SIII) was higher than that after the 57% ammonium sulfate cut (SIV). How do you explain this?

4. Investigate some of the literature cited in Additional Reading to determine how it is believed that salts and organic solvents cause the precipitation of proteins.

Additional Reading

F. B. Armstrong, *Biochemistry,* 2nd ed., Chap. 9. Oxford University Press, New York, 1983.

T. G. Cooper, *The Tools of Biochemistry.* Wiley, New York, 1977.

A. G. Gornall, C. J. Bardawill, and M. M. David, "Determination of Serum Proteins by Means of the Biuret Reaction," *J. Biol. Chem.* 177:751 (1949).

B. K. Joyce and S. Grisolia, "Purification and Properties of a Nonspecific Acid Phosphatase from Wheat Germ," *J. Biol. Chem.* 235:2278 (1960).

D. C. Wharton and R. E. McCarty, *Experiments and Methods in Biochemistry.* Macmillan, New York, 1972.

5 PROTEINS: THE KINETIC PROPERTIES OF WHEAT GERM ACID PHOSPHATASE

Enzymes are a specific class of proteins that catalyze the myriad biochemical reactions of the living cell. The precise number of enzymes in any given cell is not known but must be well into the thousands considering the vast number of reactions in the metabolic pathways. Although the details are not all worked out, it appears that an enzyme functions by forming a complex with the substrate molecule, thereby lowering the activation energy necessary to convert substrate to product. The sequence of events for a simple enzymatic reaction involving a single substrate and a single product is illustrated in Figure 5-1. Since it acts as a catalyst, the enzyme is unchanged by the reaction, its role being to speed up a process that is thermodynamically possible.

The speed or rate of the enzyme-catalyzed reaction can be determined by measuring the amount of substrate used up or product generated per unit time. This rate is called reaction velocity (v) and is most commonly expressed in terms of micromoles of product made per minute (μmoles P/minute). Since the rate of enzymatic reactions is of importance to the overall economy of the cell, it is of real significance that the biochemist learn as much as possible about the factors that affect the velocity of these reactions.

Enzyme Concentration

When all other factors (temperature, substrate concentration, etc.) are held constant and an enzymatic reaction is performed at progressively higher concentrations of enzyme, a characteristic pattern emerges (Figure 5-2). At low enzyme concentrations, the relationship is a linear one such that each higher level of enzyme yields a correspondingly higher rate of reaction. This relationship does not hold at the highest enzyme concentrations, however, because the enzyme molecules overwhelm the substrate molecules and no further velocity increases are observed. Thus, the linear portion of the graph illustrated (from O to 4 mg/ml) delimits the range of enzyme concentrations that yield valid results at the given substrate concentration.

Substrate Concentration

As might reasonably be expected, there is also a direct correlation between substrate concentration and reaction velocity. As the substrate level is increased, the velocity increases in a hyperbolic fashion (Figure 5-3). The cause for this was first elaborated by Michaelis and Menten. The shape of this curve is best explained on the basis of the degree to which substrate molecules engage the active site(s) of the enzyme. At low substrate concentration, there are many more active sites than substrate molecules; consequently they are processed with little delay, and with each increase in substrate level there is a dramatic increase in velocity. There is eventually a point, however, at which the number of substrate molecules exceeds the number of active sites; at that point the enzyme is saturated and added substrate has little effect on the reaction velocity. The mathematical expression for this curve is called

$$E + S \, \rightleftharpoons \, E:S \, \rightleftharpoons \, E + P$$

Fig. 5-1. A simple enzymatic reaction, wherein E represents the enzyme, S the substrate, P the product, and E:S the enzyme-substrate complex.

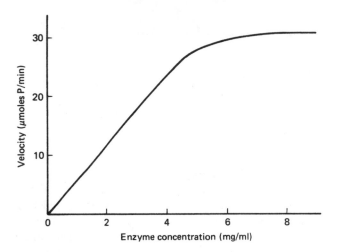

Fig. 5-2. The relationship between enzyme concentration and reaction velocity.

Fig. 5-3. The effect of increasing substrate concentration on reaction velocity. This is an example of classic Michaelis-Menten enzyme kinetics.

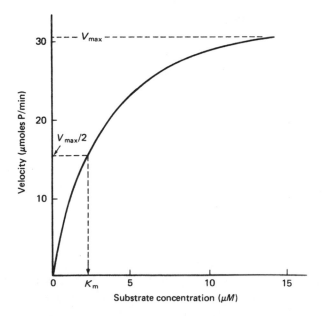

the Michaelis-Menten equation, wherein v equals the initial velocity, V_{max} equals the maximum initial velocity, [S] is the molar concentration of the substrate, and K_m is the Michaelis constant:

$$v = \frac{V_{max}[S]}{K_m + [S]}$$

Approximate values for V_{max} and K_m can be determined graphically from the Michaelis-Menten curve (Figure 5-3). V_{max} is the maximum velocity attained under the prevailing conditions (32 μmoles P/minute for the case illustrated in Figure 5-3), and K_m is the substrate concentration at one half V_{max} ($V_{max}/2$)—in this case, about 2.3 μM. The K_m is useful because it indicates the affinity of the enzyme for the substrate; a low K_m indicating a high attraction between reaction components, and a higher K_m is evidence of a low affinity.

More precise values for K_m and V_{max} may be obtained if the same experimental data are displayed graphically in the double-reciprocal plot of Lineweaver and Burke (Figure 5-4). The reciprocal of the velocity ($1/v$) plotted against the reciprocal of the substrate concentration ($1/[S]$) yields a straight line with slope K_m/V_{max}.

V_{max} is equal to the reciprocal of the y intercept (35.7 μmoles P/minute in this case), and Km is equal to the negative reciprocal of the x intercept (3.13 μM substrate). Since these results for K_m and V_{max} are determined from the x and y intercepts of a straight line, they are considered to be more reliable than those obtained by approximation from the Michaelis-Menten curve (Figure 5-3). The equation for the Lineweaver-Burke plot is obtained by taking the reciprocal of both sides of the Michaelis-Menten equation:

$$\frac{1}{v} = \frac{K_m}{V_{max}} \times \frac{1}{[S]} + \frac{1}{V_{max}}$$

The information accumulated from enzyme and substrate concentration trials is extremely useful in designing subsequent experiments. One can now set up assay conditions with a level of substrate that is not limiting and a level of enzyme that yields a linear response at that substrate concentration. For the hypothetical enzymatic reaction described thus far, ideal assay conditions might include a substrate concentration of 15.0 μM and an enzyme concentration of 4.0 mg/ml.

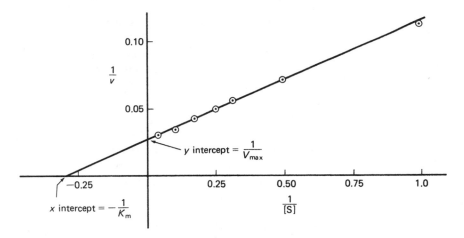

Fig. 5-4. The data presented in Figure 5–3 plotted in the double-reciprocal fashion of Lineweaver and Burke.

Enzyme Inhibitors

There exist a number of molecular species which, in the presence of an enzyme and its substrate, have the effect of binding to the enzyme (or to the enzyme-substrate complex) and totally or partially inhibiting the reaction. In those cases where the binding is irreversible, the reaction is inalterably inhibited and not subject to kinetic analysis. If the binding is reversible, however, the specific type of inhibition can be determined by kinetic analysis.

The three types of inhibition that can be clearly distinguished in this manner are competitive, noncompetitive, and uncompetitive. Experimentally, these are distinguished by performing the enzymatic reaction in the presence of a constant amount of the inhibitor at ever-increasing concentrations of the substrate. When the inhibited reaction is compared with the normal reaction using the graphic analyses of Michaelis and Menten or Lineweaver and Burke, the type of inhibition is clearly indicated. In the case of competitive inhibition, high substrate concentrations wipe out the inhibitory effect and the V_{max} for the inhibited reaction is identical to that for the uninhibited reaction (Figure 5-5).

The K_m of the inhibited reaction (K_i) is significantly higher than that of the reaction run in the absence of inhibitor, which indicates an apparent decrease in the affinity of the enzyme for its substrate. Noncompetitive inhibition yields the curve indicated in Figure 5-6, with

a lower V_{max} and a K_m (K_i) identical to that of the reaction in the absence of inhibitor.

Uncompetitive inhibition is characterized by a lower V_{max}, a higher K_m, and a Michaelis-Menten curve quite similar to that of noncompetitive inhibition. The best

Fig. 5-5. A Michaelis-Menten plot for an enzymatic reaction performed in the presence (– · – · –) and absence (——) of a competitive inhibitor.

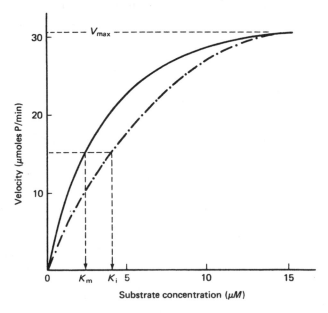

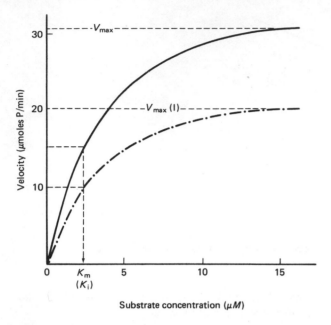

Fig. 5-6. A Michaelis-Menten plot for an enzymatic reaction performed in the presence (–·–·–) and absence (——) of a noncompetitive inhibitor.

way to distinguish the three types of inhibition graphically is to use the Lineweaver-Burke plot (Figure 5-7). Notice that in the case of uncompetitive inhibition the slope of the inhibited curve (K_m/V_{max}) is the same as that of the noninhibited curve, whereas in the other two types of inhibition, the slope of the inhibited plot is greater.

In summary, it is relatively simple to distinguish the three types of reversible inhibition by comparing the Michaelis-Menten and Lineweaver-Burke kinetics in the presence and absence of the inhibitor.

Hydrogen Ion Concentration

As mentioned earlier, enzymes are extremely sensitive to variations in pH. A change in the hydrogen ion concentration alters the charge on the amino acids of the protein, which alters the attractive forces governing the three-dimensional shape of the enzyme. Such a conformational change in the enzyme is usually reflected in the ability of the enzyme to catalyze its reaction. A typical pH curve (Figure 5-8) exhibits a peak of enzymatic activity, termed the pH optimum, and a rather precipi-

Fig. 5-7. A Lineweaver-Burke plot of an enzymatic reaction performed in the absence of inhibitor (——) and in the presence of competitive (– – – –), noncompetitive (–·–·–), and uncompetitive (· · · ·) inhibitors.

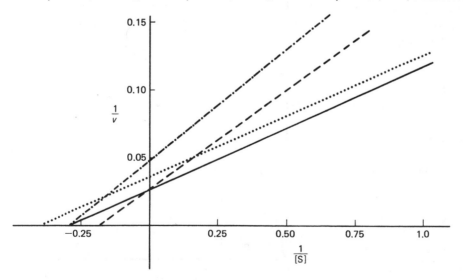

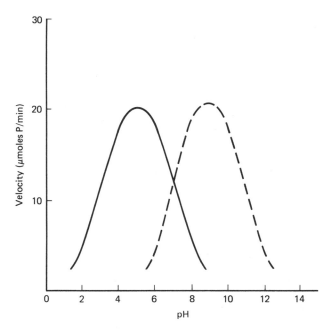

Fig. 5-8. The effect of different pH values on the reaction velocity of two different enzymes. One of the enzymes (———) has an acid pH optimum (5.0) and the other a basic pH optimum (9.0).

tous decline in activity on either side of the peak. The two curves depicted in Figure 5-8 are typical of enzymes such as acid phosphatase and alkaline phosphatase, which both catalyze the same reaction but differ in pH optimum. The sharp decline in enzymatic activity on either side of the optimum underscores the importance of maintaining an appropriately buffered solution when working with living systems *in vitro.*

Temperature

Temperature also has a profound effect on enzymatic reaction rate. At low temperatures, the enzyme is relatively inactive, which explains why ice-bath temperatures are often used in the isolation of proteins and other macromolecules sensitive to hydrolytic enzymes. At higher temperatures, there is a concomitant increase in reaction rate resulting from increased molecular motion, which continues until the optimum temperature is attained. For the majority of enzymes, activity rapidly

declines beyond that point as a result of heat denaturation of the protein (Figure 5-9).

Acid Phosphatase

The acid phosphatase you previously isolated from wheat germ will now be the object of detailed kinetic analysis. The series of experiments is designed to determine the effect of various factors (temperature, substrate concentration, etc.) on the velocity of the acid phosphatase reaction. There are five separate experiments, and you are expected to complete all of them during the next two laboratory periods. It is suggested that the experiments be performed in the following sequence:

Day 1. Time course
       Substrate concentration
       Phosphate inhibition
Day 2. Enzyme concentration
       Temperature effects

For all of these experiments, use the acid phosphatase assay described on pp. 30 to 31. Under acid con-

Fig. 5-9. The effect of different temperatures on enzymatic reaction velocity. The optimum temperature for this reaction is 30°C.

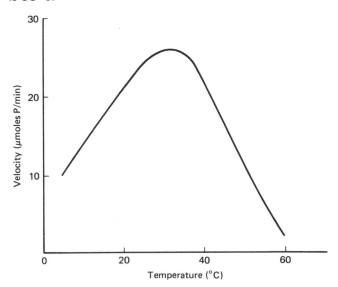

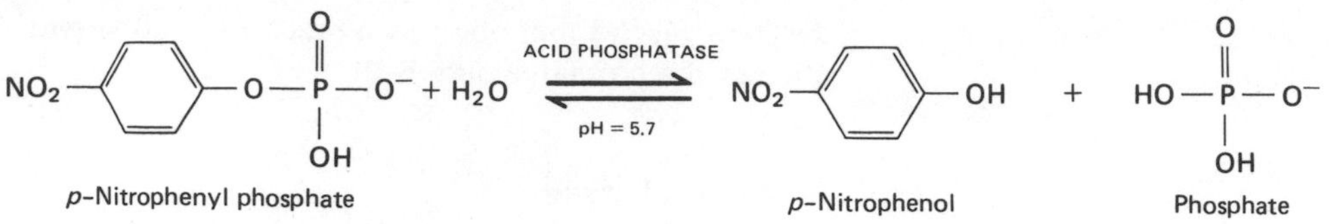

Fig. 5-10. The hydrolysis of *p*-nitrophenyl phosphate by acid phosphatase.

ditions, the enzyme catalyzes the hydrolysis of *p*-nitrophenyl phosphate to inorganic phosphate and *p*-nitrophenol (Figure 5-10). If base is added to the mixture after the completion of the reaction, the *p*-nitrophenol is converted to a colored form which absorbs light at 405 nm. Assuming a path length of 1.0 cm and an extinction coefficient of 18.8×10^3 liter mole^{-1}cm^{-1}, you can use the absorbance at 405 nm to calculate the number of micromoles of *p*-nitrophenol released. Since this is a fixed-time assay that is stopped after 5 minutes, the velocity of the reaction (μmoles of *p*-nitrophenol/minute) can be easily computed.

Each of the following experiments is done in fundamentally the same fashion, except that one factor, such as temperature or enzyme concentration, is altered to determine the effect of that parameter. Each experiment must be performed a total of three times to provide enough data for valid conclusions.

The basic assay is performed in the following manner:

1. Prepare and label the required number of reaction tubes, and to each of them add 0.5 ml of 1.0 *M* sodium acetate buffer (pH 5.7), 0.5 ml of 0.1 *M* MgCl$_2$, 0.5 ml of 0.05 *M* *p*-nitrophenyl phosphate, and 3.3 ml of distilled water.
2. Add 0.2 ml of the appropriately diluted* enzyme solution and start a stopwatch.
3. Incubate the reaction tubes at 37°C for 5 minutes.
4. Stop the reaction by adding 2.5 ml of 0.5 *M* KOH.
5. If a cloudy precipitate forms, it should be removed by a brief centrifugation at top speed in a desktop clinical centrifuge.
6. Set the colorimeter at 405 nm, adjust to zero absorbance with a blank tube, and determine the A_{405} of all assay tubes. The blank tube will differ from experiment to experiment, depending on conditions. In some cases, it will be the reaction tube lacking enzyme; in others, it will be the reaction stopped immediately after the addition of enzyme.

Kinetic Properties I.
Time Course of the Reaction

Introduction
The assay you are using is called a fixed-time assay because the reaction is stopped after 5 minutes and the velocity is calculated assuming that the relationship between product yield and time has been linear throughout. The object of this exercise is to demonstrate the validity of that assumption. Set up a series of identical enzyme reaction tubes, each of which is allowed to incubate for a different period of time (zero through 30 minutes). The results should indicate how long the reaction is linear under the given conditions of substrate and enzyme concentration.

Procedure
1. Prepare a series of seven reaction tubes labeled 0 through 30 minutes at 5-minute intervals (0, 5, 10, . . . minutes).
2. To each of these tubes add 0.5 ml of 1.0 *M* sodium acetate buffer (pH 5.7), 0.5 ml of 0.1 *M* MgCl$_2$, 0.5 ml of 0.05 *M* *p*-nitrophenyl phosphate, and 3.3 ml of distilled water.
3. Place all the tubes in a test tube rack situated in a water bath maintained at 37°C and let the temperature equilibrate for 5 minutes.
4. Add 0.2 ml of the enzyme (appropriately diluted acid phosphatase) to the tube marked 0 minutes

*An appropriate dilution of acid phosphatase yields an A_{405} of about 0.3 in 5 minutes under these assay conditions.

and immediately stop the reaction by adding 2.5 ml of 0.5 M KOH. This zero-time tube will serve as the blank against which all the others will be compared.

5. Add 0.2 ml of the enzyme to the tube marked 5 minutes, mix, start the stopwatch, and let the reaction proceed for 5 minutes before adding the KOH to terminate the reaction.

6. Run all of the other reaction tubes in exactly the same fashion with the exception that each successive tube will be incubated for 5 minutes longer than the previous one (total reaction times to equal 0, 5, 10, . . . 30 minutes). An efficient way to do this is to start each reaction at 2-minute intervals, keeping an eye on the stopwatch and stopping each of the reactions at the appropriate time. It is helpful to prepare a schedule of events (Table 5-1) before you begin.

7. After all the reactions have been terminated, determine the absorbance at 405 nm for each sample. The zero-time sample should be used as the blank.

8. Repeat this experiment two more times.

Calculations and Questions

1. Use the extinction coefficient (18.8×10^3 liter mole^{-1} cm^{-1}) for p-nitrophenol to calculate the micromoles of product released at each time point.

2. Average the data for all three trials and prepare a graph, plotting μmoles of p-nitrophenol released against time.

3. Is the time course linear throughout? If the time course is not linear, what are some factors that might contribute to the changed velocity at longer time periods?

Table 5-1.
Schedule for time course experiment

Total incubation time (min)	Clock time (min)	
	Start reaction (add enzyme)	Stop reaction (add KOH)
0	0	0
5	0	5
10	2	12
15	4	19
20	6	26
25	8	33
30	10	40

4. Is the 5-minute fixed-time assay valid for acid phosphatase? If not, how should it be changed?

5. Determine the initial velocity (v) for the acid phosphatase reaction from the slope of the linear part of the graph.

Kinetic Properties II.
The Effect of Different Substrate Concentrations on Reaction Velocity

Introduction
The object of this exercise is to demonstrate the effect of performing the standard 5-minute assay in the presence of substrate concentrations ranging from 0 to 5.0 mM p-nitrophenyl phosphate. The results should provide classic Michaelis-Menten data from which approximations of V_{max} and K_m can be made. Double-reciprocal plots of the same data should be done to arrive at even more exact values for K_m and V_{max}.

Procedure
1. Prepare a series of substrate dilutions according to the protocol outlined in Table 5-2. The amount of each sample is sufficient to do both this experiment and the next (Kinetic Properties III), which requires the same range of substrate concentrations. Do not discard these samples until both experiments have been completed.

2. Set up eight assay tubes labeled according to the various substrate concentrations. To each of these tubes add 0.5 ml of 1.0 M sodium acetate buffer

Table 5-2.
Protocol for the dilution of substrate (P-nitrophenyl phosphate)

Tube	0.05 M PNPP* (ml)	Distilled water (ml)	Concentration of dilute PNPP (M)
A	0	5.0	0
B	0.05	4.95	0.0005
C	0.10	4.90	0.0010
D	0.25	4.75	0.0025
E	0.50	4.50	0.005
F	1.00	4.0	0.010
G	2.5	2.5	0.025
H	5.0	0	0.050

*PNPP = p-nitrophenyl phosphate

(pH 5.7), 0.5 ml of 0.1 M MgCl$_2$, and 3.3 ml of distilled water. To each tube add 0.5 ml of the correspondingly diluted substrate (p-nitrophenyl phosphate).

3. Place the tubes in a test tube rack situated in a 37°C water bath and let stand for 5 minutes.
4. Initiate each assay at 2-minute intervals by adding 0.2 ml of the enzyme, run each reaction for 5 minutes, and stop it by adding 2.5 ml of 0.5 M KOH. Note that the final concentration of substrate in each reaction tube during the assay is 0, 0.05, 0.10, 0.25, 0.50, 1.0, 2.5, and 5.0 mM p-nitrophenyl phosphate.
5. Using a colorimeter adjusted to 405 nm, determine the absorbance for each reaction mixture. The tube containing no substrate should be used as the blank.
6. Repeat this experiment two more times.

Calculations and Questions
1. Determine the amount of p-nitrophenol produced in 5 minutes for each substrate concentration.
2. Average the data for all three trials and calculate the average velocity (μmoles of p-nitrophenol/minute) for each substrate concentration.
3. Plot velocity against substrate concentration (mmoles p-nitrophenyl phosphate) in the standard manner of Michaelis and Menten. Determine V_{max} and K_m for acid phosphatase.
4. Calculate the reciprocals of velocity ($1/v$) and substrate concentration ($1/[S]$) and present these data as a table.
5. Prepare the double-reciprocal plot of Lineweaver and Burke and determine the K_m and V_{max} from the x and y intercepts.
6. Investigate the literature and see how your values for K_m and V_{max} compare with published ones for acid phosphatase.

Kinetic Properties III.
The Inhibition of Acid Phosphatase by Inorganic Phosphate

Introduction
Inorganic phosphate (Pi) is an inhibitor of acid phosphatase, and it is your task to determine whether it is a competitive, noncompetitive, or uncompetitive inhibitor. This experiment must be done in conjunction (preferably the same day) with the previous one because the kinetics for the uninhibited reactions must be compared with those of reactions run in the presence of the inhibitor. The setup is basically the same as in the previous experiment (Kinetic Properties II) except that a constant amount of phosphate (1.0 mM K$_2$HPO$_4$) will be present in each reaction tube. Run the reactions as before and compare Michaelis-Menten and Lineweaver-Burke plots in the presence (Kinetic Properties III) and absence (Kinetic Properties II) of the inhibitor. Determinations of V_{max} and K_m will determine the specific mode of inhibition.

Procedure
1. Use the same set of substrate dilutions prepared for Kinetic Properties II (0 to 0.05 M p-nitrophenyl phosphate).
2. Prepare eight reaction tubes labeled in accordance with the substrate concentrations to be used. To each tube add 0.5 ml of 1.0 M sodium acetate buffer (pH 5.7), 0.5 ml of 0.1 M MgCl$_2$, 2.3 ml of distilled water, and 1.0 ml of 0.005 M K$_2$HPO$_4$. To each tube add 0.5 ml of the appropriate diluted substrate (p-nitrophenyl phosphate). Note that each tube contains a different substrate concentration and the identical inhibitor concentration.
3. Place the tubes in a 37°C water bath for 5 minutes.
4. Begin the reaction in each assay tube at 2-minute intervals by adding 0.2 ml of the enzyme, let the reactions proceed for 5 minutes, and then stop them by adding 2.5 ml of 0.5 M KOH.
5. Determine the absorbance at 405 nm for each sample, using the first tube (0 mM p-nitrophenyl phosphate) as the blank.
6. Repeat the experiment two more times.

Calculations and Questions
1. Repeat the Calculations and questions (items 1 to 5) described at the end of Kinetic Properties II.
2. Prepare Michaelis-Menten and Lineweaver-Burke plots that compare the inhibited reaction with the uninhibited reaction.
3. Determine the K_m (K_i) and V_{max} in the presence of phosphate.

4. Is phosphate a competitive, noncompetitive, or un-competitive inhibitor? Justify your answer.
5. Investigate the literature to determine how your results compare with those of previous workers.
6. What do you think would happen if you ran this same experiment substituting ATP for the inorganic phosphate?

Kinetic Properties IV.
The Effect of Different Enzyme Concentrations on Reaction Velocity

Introduction
This exercise illustrates the effect of increasing enzyme concentrations on reaction rate. You will perform a series of 5-minute assays, in which a different enzyme concentration is added each time the reaction is initiated. The results should indicate the range of enzyme concentrations that yield a linear response.

Procedure
1. Prepare a series of enzyme dilutions according to the protocol in Table 5-3. Notice that the enzyme is to be diluted with a solution of bovine serum albumin (BSA) at a concentration of 1 mg/ml. This added protein protects the enzyme against denaturation at low concentration. You must fill in the last column of Table 5-3 (enzyme concentration) based upon the actual concentration of your own sample of enzyme.
2. Prepare nine reaction tubes labeled according to the various enzyme concentrations (units/ml) and to each of these add 0.5 ml of 1.0 M sodium acetate buffer (pH 5.7), 0.5 ml of 0.1 M $MgCl_2$, 0.5 ml of 0.05 M p-nitrophenyl phosphate, and 3.3 ml of distilled water.
3. Place all the tubes in a test tube rack situated in a 37°C water bath and let the temperature equilibrate for 5 minutes.
4. Using a time schedule patterned after the one described in Table 5-1, start the reactions at 2-minute intervals by adding 0.2 ml of the different enzyme concentrations to each of the corresponding reaction tubes. Stop each reaction after 5 minutes by adding 2.5 ml of 0.5 M KOH.
5. Determine the absorbance for each reaction mixture

Table 5-3.
Protocol for the dilution of acid phosphatase

Tube	Undiluted enzyme (ml)	BSA (ml)	Dilution factor	Enzyme conc. (units/ml)
A	0	1.00	—	
B	0.01	0.99	1:100	
C	0.02	0.98	1:50	
D	0.04	0.96	1:25	
E	0.10	0.90	1:10	
F	0.20	0.80	1:5	
G	0.40	0.60	2:5	
H	0.80	0.20	4:5	
I	1.00	0	none	

at 405 nm using the tube containing no enzyme as the blank.
6. Repeat the experiment two more times.

Calculations and Questions
1. Use the extinction coefficient for p-nitrophenol to determine the micromoles of product produced in 5 minutes at each of the enzyme concentrations.
2. Average your data for the three trials and calculate the average reaction velocity (μmoles p-nitrophenol/minute) for each enzyme concentration. Present these data in tabular form.
3. Plot velocity against enzyme concentration (units/ml). Describe the shape of this curve and discuss the reasons for its shape.
4. What is the valid range of enzyme concentrations for the acid phosphatase assay?

Kinetic Properties V.
Effects of Different Temperatures on Reaction Velocity

Introduction
As is the case with all chemical reactions, enzymatic reactions are sensitive to changes in temperature. This

exercise will demonstrate that phenomenon in two different ways: (1) by showing the effect on enzyme stability of enzyme preincubation at a range of temperatures and (2) by showing the effect of different temperatures on the rate of the enzymatic reactions. For the first experiment, samples of the enzyme are maintained at eight predetermined temperatures (0, 10, 20, 30, 37, 50, 80, and 100°C) for 30 minutes, cooled in an ice bath, and then used in the standard 5-minute assay run at 37°C. The second experiment employs the standard stock of enzyme (no previous temperature treatment) in regular 5-minute assays performed at the eight different temperatures.

Procedure

1. One factor that is critical to the success of these experiments is the preparation and maintenance of water baths at different temperatures. Once a bath has been adjusted, its temperature should be continuously monitored and all temperature changes recorded. Baths prepared in the manner described in Table 5-4 can be shared by several laboratory pairs.

2. The effect on enzyme stability of enzyme preincubation at different temperatures
   a. Place a small sample (1.0 ml) of the enzyme in each of the water baths.
   b. After 30 minutes, return the samples to an ice bath.
   c. Prepare eight labeled enzyme assay tubes (one for each temperature) and add to each 0.5 ml of 1.0 M sodium acetate buffer (pH 5.7), 0.5 ml of 0.1 M MgCl$_2$, 0.5 ml of 0.05 M p-nitrophenyl phosphate, and 3.3 ml of distilled water.
   d. Place the tubes in a test tube rack situated in a 37°C water bath and let them equilibrate for 5 minutes.
   e. Start each of the enzyme reactions at 2-minute intervals by adding 0.2 ml of the appropriately treated enzyme to the corresponding reaction mixture. Let each reaction proceed at 37°C for 5 minutes, terminating it by the addition of 2.5 ml of 0.5 M KOH.
   f. Prepare a blank assay tube containing all the reaction components except the enzyme and incubate it at 37°C for 5 minutes along with the other tubes. After the KOH has been added to this tube, it may be used to zero the colorimeter at 405 nm.
   g. Determine the absorbance of each reaction mixture.
   h. Repeat the experiment two more times.

3. The effect of performing the reaction at different temperatures on the rate of the reaction
   a. Label four assay tubes (A, B, C, D) and into each of them pipette 0.5 ml of 1.0 M sodium acetate buffer (pH 5.7), 0.5 ml of 0.1 M MgCl$_2$, 0.5 ml of 0.05 M p-nitrophenyl phosphate, and 3.3 ml of distilled water.
   b. Place the tubes in a water bath maintained at 0 to 4°C and let the temperature equilibrate for 5 minutes.
   c. Add 0.2 ml of enzyme to tubes B, C, and D at 2-minute intervals and allow each reaction to proceed for 5 minutes before stopping it with the addition of 2.5 ml of 0.5 M KOH. Tube A, which serves as a reagent blank, should be treated in the same fashion except that 0.2 ml of distilled water should be added to the reaction mixture instead of enzyme.
   d. Place the tubes in a test tube rack at room temperature.
   e. Repeat steps a through d using all the water bath temperatures described in Table 5-4. When all of the reaction mixtures have returned to room temperature, determine the absorbance at 405 nm of each experimental tube against its own blank tube (A).

Table 5-4.
Procedure for the preparation of water baths of different temperatures

Desired temperature (°C)	Method of preparation
0–4	Ice plus tap water in an ice bucket
10	Tap water and ice
20	Tap water at room temperature
30	Thermostatted water bath
37	Thermostatted water bath
50	Thermostatted water bath
80	Hot tap water
100	Boiling water bath

Calculations and Questions

1. Convert absorbance data to velocity data and compute the average velocity for all reactions at each temperature.

2. Report the data for the first experiment in a tabular format illustrating enzymatic activity remaining after enzyme preincubation at various temperatures. This is a measure of enzyme stability.

   a. Does it appear that in any case the enzymatic activity was increased or decreased by the temperature treatments?

   b. How do you explain these results?

   c. If you planned to store the enzyme for a prolonged period of time, what temperature would you suggest? Why?

3. Report the results of the second experiment in the form of a graph illustrating the effect of different temperatures on the rate of the reaction.

   a. Does this graph conform with your expectations? If not, why not?

   b. What is the optimum temperature for acid phosphatase?

   c. Why do you think it was necessary to prepare individual reagent blanks to be run at each of the temperatures?

4. Discuss the long- and short-term effects of different temperatures on acid phosphatase.

Additional Reading

M. A. Andersch, and A. J. Szczypinski, "Use of *p*-Nitrophenyl Phosphate as the Substrate in the Determination of Serum Acid Phosphatase," *Am J. Clin. Pathol.* **17:** 571 (1947).

F. B. Armstrong, *Biochemistry,* 2nd ed., Chap. 9. Oxford University Press, New York, 1983.

J. M. Clark, Jr. and R. W. Switzer, *Experimental Biochemistry,* 2nd ed. Freeman, San Francisco, 1977.

E. C. Conn, and P. K. Stumpf, *Outlines of Biochemistry,* 4th ed. Wiley, New York, 1976.

T. G. Cooper, *The Tools of Biochemistry.* Wiley, New York, 1977.

M. Dixon, and E. C. Webb, *Enzymes,* 2nd ed. Academic Press, New York, 1964.

V. P. Hollander, "Acid Phosphatases," in *The Enzymes,* 3rd ed., Vol. IV. Academic Press, New York, 1971.

B. K. Joyce, and S. Grisolia, "Purification and Properties of a Nonspecific Acid Phosphatase from Wheat Germ," *J. Biol. Chem* **235:**2278 (1960).

A. L. Lehninger, *Biochemistry,* 2nd ed. Worth, New York, 1975.

A. J. Sommer, "The Determination of Acid and Alkaline Phosphatase Using *p*-Nitrophenyl Phosphate as Substrate," *Am. J. Med. Tech.* **20:**244 (1954).

D. C. Wharton, and R. E. McCarty, *Experiments and Methods in Biochemistry.* Macmillan, New York, 1972.

6 CARBOHYDRATES: THE ANALYSIS OF GLYCOGEN

The carbohydrates include the simple sugars and the polymers thereof. They are most rigorously defined as polyhydroxy aldehydes or ketones and their derivatives. These versatile molecules serve many different functions in the cell, two of which stand out as being of primary importance: (1) they are readily metabolized and serve as a rich source of ATP (adenosine triphosphate) and (2) polymers of simple sugars serve as structural components of cell walls. Carbohydrates constitute a relatively small proportion of the total cell mass (about 3 per cent), but they are easily the most abundant biomolecule on earth. The reason for this will become evident shortly.

Many different organisms can manufacture carbohydrates, but green plants, through the process of photosynthesis, synthesize the vast majority of these compounds. Plants use sunlight, atmospheric carbon dioxide, and water to produce simple sugars, such as glucose. The glucose may be used directly as an energy source, stored in the form of starch, or used to build cellulose cell walls. Nonphotosynthetic organisms generally obtain carbohydrates either directly or indirectly from plants.

Monosaccharides

Simple sugars such as ribose or glucose are termed monosaccharides, and polymers of sugars are called oligosaccharides and polysaccharides. Oligosaccharides contain from two to ten simple sugars linked by glycosidic bonds, and polysaccharides contain hundreds or thousands of sugars joined in this fashion.

Glucose (Figure 6-1) is a six-carbon monosaccharide found almost universally throughout the living world. It can be used by practically all organisms as a source of energy and as a building-block molecule for such polysaccharides as starch, glycogen, and cellulose.

Polysaccharides

Cellulose is the major structural component of plant cell walls and is probably the most abundant natural polymer in the world. It is composed of glucose molecules joined by β (1→4) glycosidic bonds forming a high-molecular-weight (several million daltons), unbranched chain (Figure 6-2). As is appropriate to such a molecule, cellulose is insoluble in water and resistant to acid and most enzymatic attack. This resistance to degradation explains why carbohydrates are so abundant. Lower organisms, such as bacteria and fungi, possess cellulases, which are enzymes capable of hydrolysing cellulose. Higher organisms, with the exception of ruminants (cows, goats, etc.), do not have cellulases and cannot use cellulose as an energy source. The ruminants harbor cellulase-producing bacteria in their gut, and these aid in the digestion of cellulose. The task then of degrading plant material falls almost exclusively to the microorganisms.

Starch is the energy-storage polymer synthesized by plants. Two distinct molecules are found in starch grains, amylose and amylopectin, both of which are made exclusively of glucose. Amylose is an unbranched chain of glucose linked by α (1→4) bonds. Amylopectin is a branched molecule in which the backbone glucoses

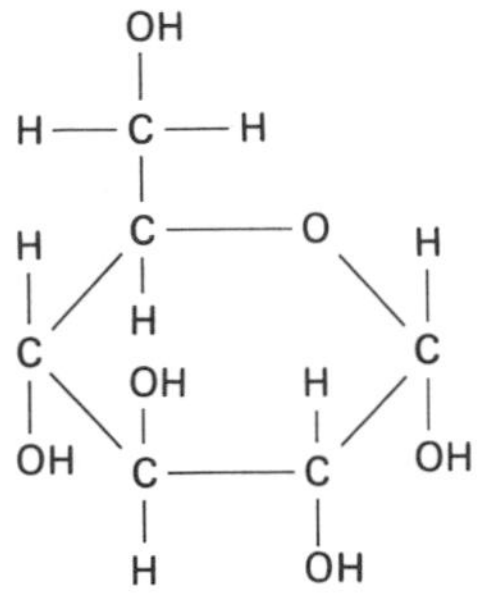

Fig. 6-1. The monosaccharide glucose.

are joined by α (1→4) bonds and the branch points arise at α (1→6) linkages (Figure 6-3).

Glycogen is the storage polymer of animals and bacteria. It is practically identical to amylopectin except that it is more highly branched. All of these storage polymers are synthesized when there is an excess of glucose in the system and degraded when there is a need for energy. In the cell, this degradation is accomplished by a variety of enzymes which specifically cleave the bonds between glucose moieties. The liberated mono- and disaccharides are further metabolized by the cell to form ATP.

The exercises described below demonstrate some of the quantitative and qualitative methods available for the analysis of carbohydrates. Nelson's method for the quantitative determination of reducing sugars demonstrates how much glucose is liberated from glycogen during acid hydrolysis. Thin layer chromatography is used qualitatively to demonstrate the difference between the reaction products of the acid hydrolysis of glycogen and the enzymatic hydrolysis of glycogen.

The three exercises should be completed in two laboratory sessions, according to the following schedule:

Day 1. Glucose analysis
Acid hydrolysis of glycogen
Day 2. Analysis of glycogen by thin layer chromatography

Carbohydrates I. Glucose Analysis

Introduction
Nelson's method is used to establish a standard curve for glucose. Samples containing accurately known concentrations of glucose are subjected to this colorimetric assay, absorbance readings recorded, and the data plotted as a standard curve. It should be borne in mind that this method is a general test for reducing sugars and does not distinguish between reducing monosaccharides (such as glucose) and reducing disaccharides (such as maltose).

Procedures
1. Preparation of glucose standards
   a. Obtain about 10 ml of 0.5 mM glucose.
   b. Transfer measured amounts of the glucose solution into two identical sets of labeled test tubes according to the protocol in Table 6-1. Add sufficient distilled water to each tube to bring the final volume up to 1.0 ml.
   c. There should now be two sets of seven tubes each containing concentrations of glucose ranging from 0 to 0.5 μmoles/ml.

Fig. 6-2. A short length of cellulose illustrating the β(1→4) linkage between glucose molecules and the unbranched chain.

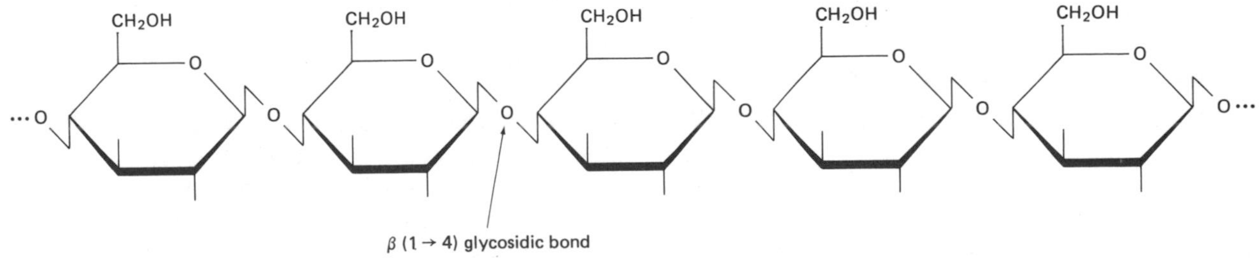

β (1 → 4) glycosidic bond

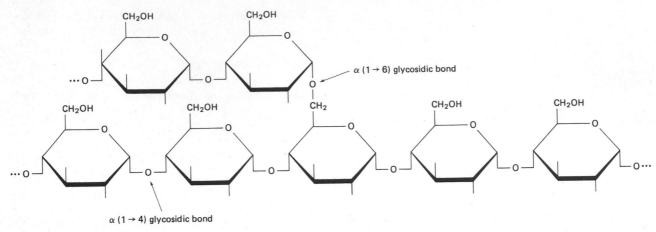

Fig. 6-3. A portion of amylopectin (or glycogen) illustrating the branching chain and the two types of glycosidic bonds.

Table 6-1.
Protocol for glucose standard curve

Tube	0.5 mM glucose (ml)	Distilled water (ml)	Glucose concentration (μmoles/ml)
1	0.0	1.0	0
2	0.1	0.9	0.05
3	0.2	0.8	0.1
4	0.4	0.6	0.2
5	0.6	0.4	0.3
6	0.8	0.2	0.4
7	1.0	0.0	0.5

2. Nelson's test for glucose
   a. Add 1.0 ml of Nelson's reagent to each tube (this must be freshly prepared each day by mixing 25 parts of Nelson's reagent A with 1 part of Nelson's reagent B).
   b. Place an aluminum foil cap on the top of each tube and heat the tubes in a vigorously boiling water bath for *exactly* 20 minutes. Cool the tubes in a beaker of cool tap water.
   c. Add 1.0 ml of the arsenomolybdate reagent to each tube, mix, and allow the tubes to stand for a few minutes.

CAUTION: The arsenomolybdate reagent contains sodium arsenate, a deadly poison, and you must not mouth pipette this reagent!

d. Make the volume of each tube up to 10.0 ml with distilled water, mix, and determine the absorbance at 510 nm.

Calculations and Questions
1. Prepare a standard curve by plotting the average absorbance at 510 nm against glucose concentration (μmoles/ml). If the curve is not linear, the procedure must be repeated.
2. From the slope of this graph, calculate the absorbance given by 1.0 μmole of glucose under these conditions. This value is called the *absorptivity index* and may be used to convert absorbance to micromoles for samples containing unknown amounts of glucose.
3. Is there any obvious reason why this assay could not be used for maltose or other simple sugars?

Carbohydrates II. Acid Hydrolysis of Glycogen

Introduction
The composition of glycogen can be demonstrated *in vitro* by totally hydrolysing it in the presence of acid. The acid hydrolysis of glycogen does not take place immediately; rather it takes place over a period of time. One can then follow the process of the reaction from beginning to end by removing small samples at prede-

termined time intervals and assaying them for glucose. Initially, little or no glucose is evident, but as the reaction progresses, increasingly more glucose is released until all of the glycogen is degraded.

In brief, the exercise is as follows: immediately after adding acid to the glycogen solution, remove a small sample of the mixture and neutralize it with potassium phosphate. This stops the hydrolysis reaction so that the sample can be set aside and assayed for glucose at a later time. The remaining glycogen-acid mixture is heated for 30 minutes, during which time additional samples are removed and neutralized for subsequent analysis. When hydrolysis is complete, all of the samples are analyzed for glucose by the Nelson method.

Procedure

1. Acid hydrolysis of glycogen
   a. Prepare a series of seven test tubes, labeled O through 30 minutes in 5-minute intervals.
   b. Pipette 2.5 ml of 1.0 M K_2HPO_4 into each tube.
   c. In a separate tube, add 4.0 ml of glycogen (8 mg/ml) to 4.0 ml of 4.0 M hydrochloric acid and mix.
   d. *Immediately* withdraw 0.5 ml of the glycogen-acid mixture and transfer it to the tube marked O minutes.
   e. Place the remainder of the glycogen-acid solution in a vigorously boiling water bath, using a foil cap to cover the top of the tube.
   f. Remove 0.5-ml samples at 5-minute intervals for 30 minutes and transfer them to the appropriately labeled tubes containing 1.0 M K_2HPO_4.
   g. Dilute each of the seven samples to 10.0 ml by adding 7.0 ml of distilled water.

2. Determination of glucose by the Nelson method
   a. Transfer 0.5 ml of each sample (O to 30 minutes) to a duplicate series of labeled test tubes.
   b. Bring the volume in each tube up to 1.0 ml by adding distilled water.
   c. Test each of the 1.0-ml samples for the presence of glucose using the Nelson's method described in Carbohydrates I (Procedures, step 2). When determining the absorbance, use one of the zero-time samples as a blank.

Calculations and Questions

1. Calculate the average absorbance (510 nm) for each time point.
2. Using the standard curve from the previous exercise (Carbohydrates I), compute the micromoles of glucose released during each time interval.
3. Present these data in the form of a graph and:
   a. Determine the initial reaction rate for glycogen hydrolysis.
   b. Determine the time required for total hydrolysis.
4. From the glucose content of a completely hydrolysed sample, determine the number of micromoles of glucose per milligram of glycogen.
5. If starch or cellulose were treated in this fashion, what would you predict for results?

Carbohydrates III. Analysis of Glycogen by Thin Layer Chromatography

Introduction

The previous exercise demonstrates that glycogen is a polymer made up of glucose subunits, which can be released by acid hydrolysis over a relatively short period of time. In the cell *(in vivo)* glycogen functions as a storehouse of glucose, which can be utilized by the cell for the generation of energy in the form of ATP. However, before the glycogen can be used by the cell it must be degraded to glucose, and since the cell does not normally function well at a low pH, the degradation cannot be accomplished by acid hydrolysis. The cell possesses a variety of enzymes (hydrolases such as α-amylases and α (1→6) glucosidases) which achieve the hydrolysis of glycogen at or near neutrality.

This exercise deals with the hydrolysis of glycogen by α-amylase (α (1→4) glucan, 4-glucanohydrolase, E.C.3.2.1.1.), an enzyme commonly found in saliva and pancreatic extracts. Human salivary α-amylase has been extensively purified and characterized. The enzyme has a molecular weight of 5.5×10^4 and has a cofactor requirement for anions (such as Cl^- or Br^-) for activity. This enzyme attacks α (1→4) linkages in the interior of the glycogen molecule, yielding a mixture of glucose, maltose, and dextrins.

A crude sample of human salivary amylase (your own) is mixed with glycogen and the resulting hydrolysate analyzed by thin layer chromatography (TLC). For purposes of comparison, an acid hydrolysate of glycogen and pure samples of glucose and maltose are also analyzed. The results of this exercise should clearly demonstrate that the products of enzymatic hydrolysis differ from those of acid hydrolysis.

Procedure

1. Preparation of TLC plates
   a. Obtain two TLC plates (20 × 5 cm) coated with silica gel.
   b. Pretreatment to facilitate sugar separation
      1) Using a shallow glass dish, soak the plates in a solution of 0.1 M KH_2PO_4 for 2 or 3 minutes.
      2) Heat the plates in an oven at 110°C for 1 hour.
      3) Cool to room temperature.
   c. Using a pencil (no ink!), gently draw a line 2 cm from the bottom of each plate and mark it off in 1-cm divisions. Label the equally spaced points according to the samples you plan to apply to the chromatograms (Figure 6-4).
2. Preparation of acid hydrolysate
   a. Add 2.0 ml of 4.0 M HCl to 2.0 ml of glycogen (100 mg/ml).
   b. Cover the tube with an aluminum foil cap and place it in a boiling water bath for 30 minutes.
   c. Remove the tube and cool it to room temperature in a cool water bath.
   d. Transfer 1.0 ml of the acid hydrolysate (AH) to a tube containing 5.0 ml of 1.0 M K_2HPO_4. The potassium phosphate should neutralize the acid solution. Check the pH with pH paper and, if necessary, adjust the pH to 7.0 by the dropwise addition of 1.0 M NaOH or 1.0 M HCl.
   e. Save this neutralized AH for analysis by TLC.
3. Preparation of the enzyme hydrolysate
   a. One member of the laboratory pair should collect approximately 2.0 ml of saliva by spitting into a beaker. Prepare a tenfold dilution of the saliva with distilled water.
   b. Prepare a test tube containing the following reaction mixture: 0.5 ml of 0.5 M potassium phosphate buffer (pH 6.9), 1.0 ml of 0.1 M NaCl, 3.0 ml of glycogen (100 mg/ml), and 4.5 ml of distilled water.

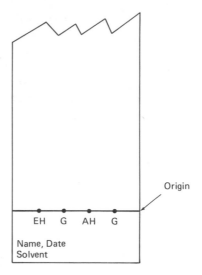

Fig. 6-4. A silica gel plate set up to run glucose (G) in parallel with the acid hydrolysate (AH) and the enzyme hydrolysate (EH).

   c. To start the enzymatic reaction, add 1.0 ml of the diluted saliva to the reaction tube, mix, and allow to incubate at room temperature for 30 minutes.
   d. This enzyme hydrolysate (EH) should be saved on ice for analysis by TLC.
4. Application of carbohydrate samples to TLC plates
   a. Using a micropipette, apply 10 μl of each sample to the origin of the chromatogram. Apply the sample in very small amounts, barely touching the silica gel long enough to let a small portion flow out of the capillary. After each application, let the spot dry before adding more liquid to the spot. Drying can be speeded by use of a hair dryer or heat gun. The individual spots should be no larger than 4 mm in diameter.
   b. Be careful not to press down on the silica gel with the capillary because it may make holes in the surface and cause uneven running of the samples.
   c. In the manner described, prepare a total of two TLC plates which compare the movement of glucose and maltose to that of the EH and the AH.
5. Chromatography and localization of carbohydrate samples

a. Place both plates in a chromatography tank filled with solvent (chloroform:glacial acetic acid:water, 30:35:5) to a level about 0.5 cm from the bottom. Cover the tank with a glass plate, place in a fume hood, and let the solvent run until the front is about 10 cm from the origin (3 to 4 hours).

b. Remove the plates from the chromatography tank and mark the position of the solvent front with a pencil. Allow the plates to air dry.

c. Determine the position of the sugars on the chromatograms as follows.

1) Spray each chromatogram with the locating solution (*p*-anisidine and phthalic acid in methanol).

2) Heat the plates to 100°C in an oven for 5 minutes. The sugars will be evident by the appearance of brown or black spots. Encircle each spot with a light pencil line.

d. Prepare an accurate, labeled sketch of each chromatogram for your notebook. The diagram should include all relevant information—solvent front, origin, etc.

Calculations and Questions

1. Calculate the R_f values for each spot on both chromatograms.

a. Measure the distance (in cm) from the origin to the solvent front and from the origin to the leading edge of each spot.

b. Use the following formula to compute the R_f value:

$$R_f = \frac{\text{distance from origin to spot}}{\text{distance from origin to solvent front}}$$

c. Enter the average R_f values for glucose, maltose, and the hydrolysed samples in a chromatography data sheet in your notebook.

2. Compare the R_f values of the known standard solutions of glucose and maltose with the products of enzyme hydrolysis and acid hydrolysis in light of your understanding of enzyme specificity.

3. Using data from the previous exercise (Carbohydrates II), calculate the total amount of hydrolysed glycogen applied to each chromatogram (AH and EH). Would you judge that analysis of sugars by TLC is more or less sensitive than the Nelson's test? Discuss the advantages and disadvantages of both methods.

4. Did you find other spots on the chromatograms that cannot be identified as glucose or maltose? What do you suspect they are?

Additional Reading

F. B. Armstrong, *Biochemistry,* 2nd ed., Chaps. 11 and 18. Oxford University Press, New York, 1983.

E. E. Conn and P. K. Stumpf, *Outlines of Biochemistry,* 4th ed. Wiley, New York, 1976.

M. Dixon and E. C. Webb, *Enzymes.* Academic Press, New York, 1964.

E. H. Fisher and E. A. Stein, "α-Amylases," in *The Enzymes,* H. Lardy and K. Myrback, eds. Academic Press, New York, 1960.

A. L. Lehninger, *Biochemistry,* 2nd ed. Worth, New York, 1975.

N. Nelson, "A Photometric Adaptation of the Somogyi Method for the Determination of Glucose," *J. Biol. Chem.* **153**:375 (1944).

W. W. Pigman, ed., *The Carbohydrates.* Academic Press, New York, 1957.

K. Randerath, *Thin Layer Chromatography.* Academic Press, New York, 1968.

G. Rendina, *Experimental Methods in Modern Biochemistry.* Saunders, Philadelphia, 1971.

D. C. Wharton and R. E. McCarty, *Experiments and Methods in Biochemistry.* Macmillan, New York, 1972.

7 LIPIDS: AN ANALYSIS OF SOME COMMON FATS AND OILS

Most biomolecules of the cell are readily categorized on the basis of a common building-block molecule or some common function. This is not the case, however, for the lipids, which are a very diverse group. These fatty molecules are most conveniently grouped on the basis of their insolubility in water and solubility in nonpolar solvents, such as chloroform, ether, and benzene. Most lipids also share the property of saponification, which is the formation of fatty acid salts upon alkaline hydrolysis. Those lipids that are saponifiable include triacylglycerols, phosphoglycerides, glycolipids, sphingolipids, and the waxes. The nonsaponifiable lipids are the terpenoids. In this chapter, the structure and role of representative lipids from each group are discussed.

Fatty Acids

As an important component of many lipids, the fatty acid is an appropriate topic for special consideration. It is best described as a long hydrocarbon chain terminating in a carboxyl group. The smallest fatty acid is acetic acid with its two-carbon chain. Very long fatty acids with upwards of 28 carbons do exist; however, those with 16 to 18 carbons predominate in the cell. Fatty acids differ from one another not only in size but also in degree of saturation. Palmitic acid is a typical medium-length, saturated fatty acid (Figure 7-1). Those with one or more double bonds in the hydrocarbon chain, such as oleic acid (Figure 7-1), are called unsaturated fatty acids.

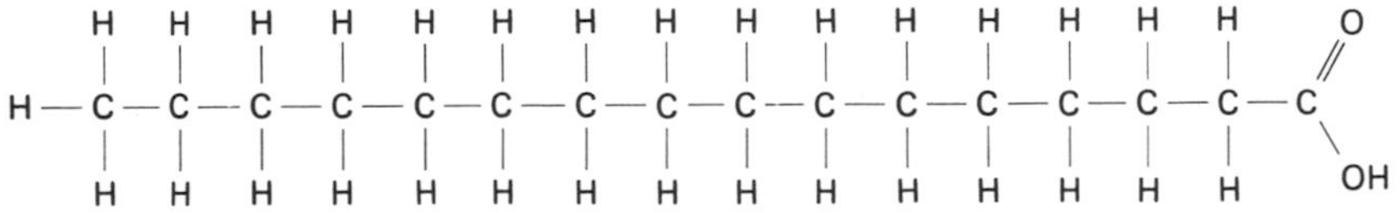

Palmitic acid

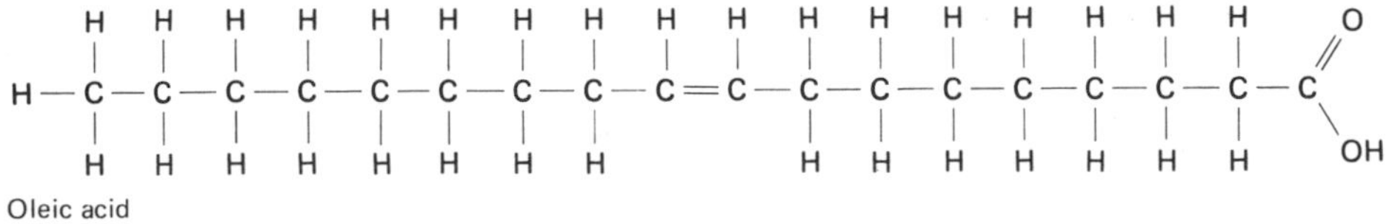

Oleic acid

Fig. 7-1. Two common medium-length fatty acids.

Some fatty acids are liquid at room temperature and others are solid. This property depends upon the size of the molecule and degree of saturation. The melting point data in Table 7-1 illustrate two points: (1) with increasing chain length the melting point increases and (2) additional double bonds have the effect of depressing the melting point.

Fatty acids are rarely found in the cell as such because they are toxic in that form. They are more typically found in combination with other chemical species, such as carbohydrates or proteins. When these fatty-acid-containing molecules are metabolized by the cell, they yield large amounts of energy in the form of ATP. Fatty acids, then, are an important source of cellular energy.

Triacylglycerols

The triacylglycerols constitute the bulk of the storage fats and oils found in plants and animals. The adipose tissues of animals are laden with these energy-rich molecules, as are the seeds of many plant species. The term *fat* is generically applied to those triacylglycerols that are solid at room temperature, and those that are liquid at room temperature are called oils.

Chemically, a triacylglycerol is a fatty acid ester of glycerol. Most naturally occurring fats are esterified at all three positions, although mono- and diacylglycerols do exist. Those fats containing but a single kind of fatty acid, such as tripalmitin (Figure 7-2), are referred to as simple triacylglycerols, and those containing two or three different fatty acids are called mixed.

There is considerable variation in the fatty acid composition of oils and fats isolated from different natural sources. Codfish oil contains more than 12 different fatty acids, whereas olive oil has only four. Whether a given triacylglycerol is solid or liquid at room temperature depends upon the properties of the constituent fatty acids. A fairly dependable generalization holds that those from plant sources have a preponderance of unsaturated fatty acids and are liquid at room temperature and those from animal tissues contain primarily saturated fatty acids and are solid at room temperature. A notable exception to this generalization is coconut oil, which contains well over 70% saturated fatty acids.

Table 7-1.
Melting point data for various fatty acids.

Saturated fatty acids of increasing size		
Name	No. of carbons	Melting point (°C)
Lauric	12	44
Palmitic	16	63
Arachidic	20	77
Lignoceric	24	86
Fatty acids having the same number of carbon atoms (18)		
Name	No. of double bonds	Melting point (°C)
Stearic	0	70
Oleic	1	13
Linoleic	2	−5
Linolenic	3	−10

Saponification of triacylglycerols yields a mixture of fatty acid salts (soaps) and glycerol (Figure 7-3). This is how soaps are manufactured from fats and lye. Soaps are amphipathic in nature and form micelles in aqueous media. This property makes them excellent cleansing agents. Addition of strong acid to a soap solution has the effect of displacing the sodium ions with hydrogen ions and causing the fatty acids to precipitate (Figure 7-4).

Phosphoglycerides

The phosphoglycerides (also called phospholipids) are structurally similar to the triacylglycerols and are an important component of cellular membranes. A typical phosphoglyceride consists of glycerol esterified at position 1 with a saturated fatty acid, position 2 with an unsaturated fatty acid, and position 3 with a phospho-

Fig. 7-2. Tripalmitin, a simple triacylglycerol.

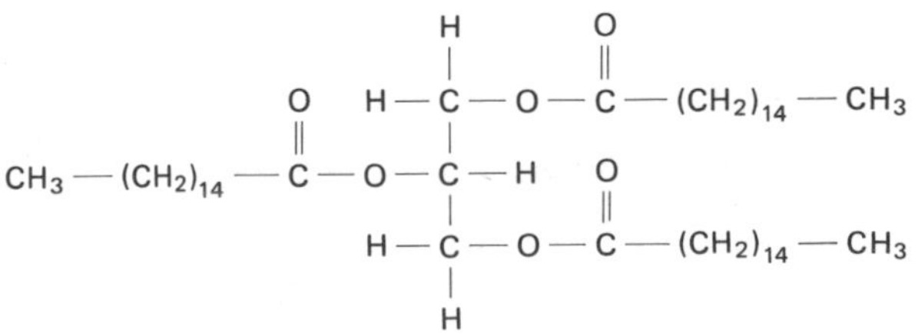

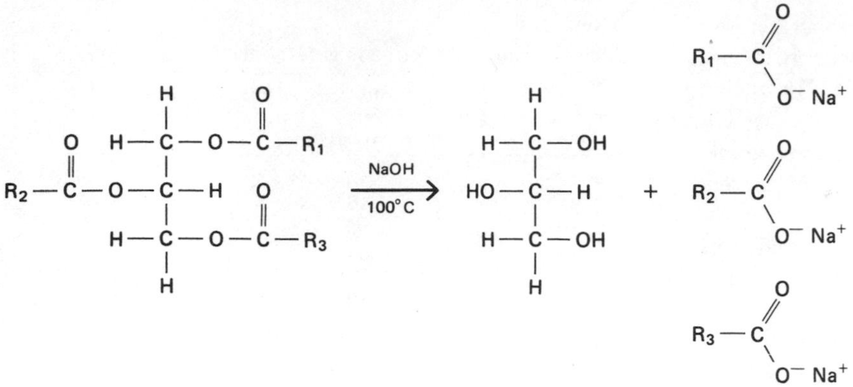

Triacylglycerol Glycerol Soaps

Fig. 7-3. Saponification, the alkaline hydrolysis of a mixed triacylglycerol. The R_1, R_2, and R_3 represent hydrocarbon chains of differing lengths and degrees of saturation.

Fig. 7-4. The precipitation of fatty acids from soap by acidification.

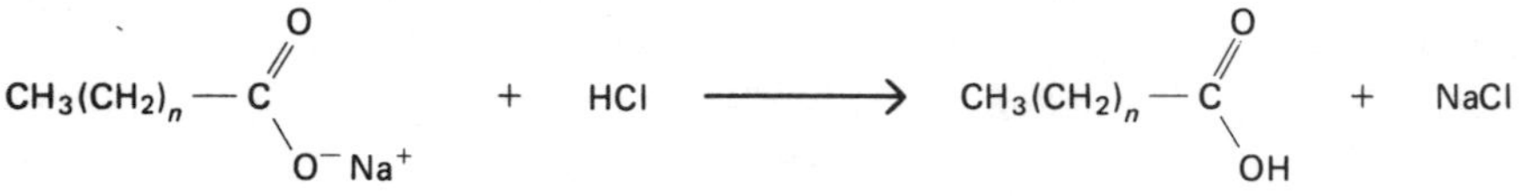

Fatty acid soap Free fatty acid

rylated alcohol group. Figure 7-5 illustrates such a molecule.

Fig. 7-5. Phosphatidyl ethanolamine, a phospholipid found in the membranes of plants, animals, and microorganisms. The first fatty acid (R_1) would probably be stearic or palmitic, and the second (R_2) is likely to be oleic.

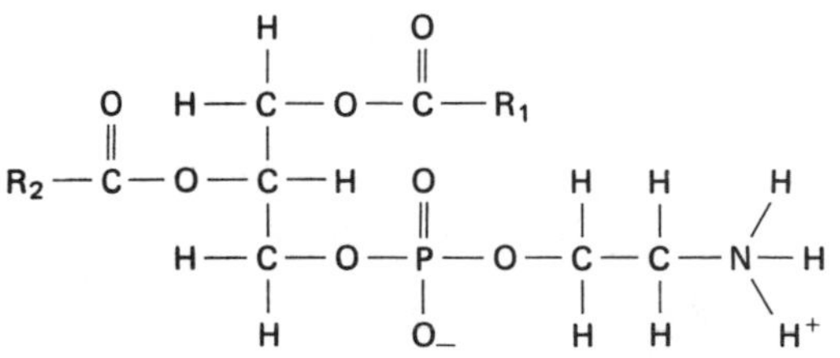

In terms of water solubility, the polar end, or head, of the phospholipid contrasts sharply with the long hydrocarbon chains. These molecules are amphipathic and are usually diagrammed as in Figure 7-6, with the hydrophilic end projecting away from the nonpolar fatty acid chains.

In aqueous media these molecules will form micelles, monomolecular films, and lipid bilayers (Figure 7-7). It is this property of phospholipids that suits them so ideally for the lamellar substructure of cellular membranes.

Glycolipids

These molecules are found mostly in the membranes of brain and nervous tissue and possess the same amphipathic properties the phosphoglycerides do. An exam-

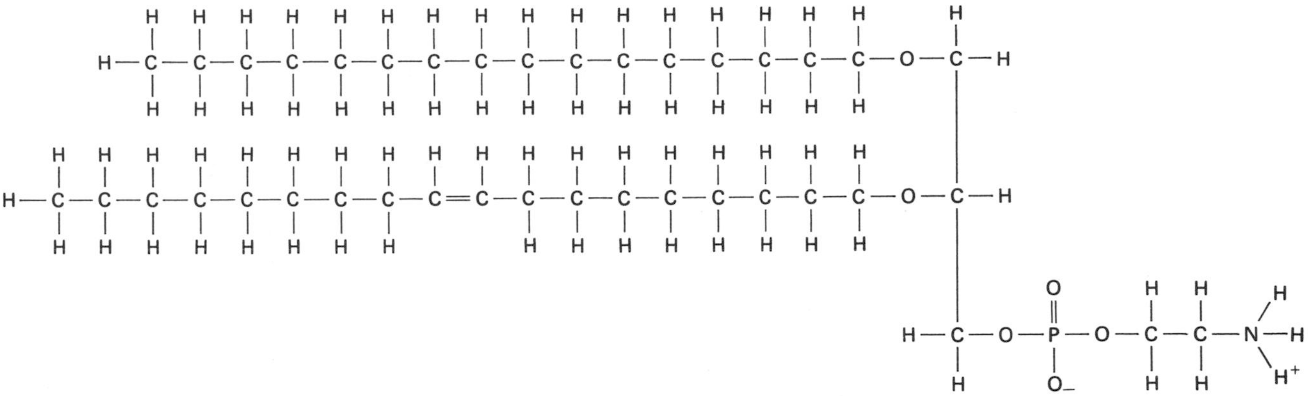

Fig. 7-6. Phosphatidyl ethanolamine as it most probably exists in an aqueous medium.

ple of a glycolipid containing two fatty acids and galactose is pictured in Figure 7-8. It is the presence of the carbohydrate moiety that gives this group its characteristic name.

Fig. 7-7. Above, a representation of a single phospholipid molecule, such as phosphatidyl ethanolamine. Below, the various associations of these amphipathic molecules with one another and with a polar solvent.

Sphingolipids

Sphingolipids, which are also found in brain and nervous tissue, contain the molecule sphingosine as well as fatty acids, phosphate, and a nitrogen-containing alcohol. Figure 7-9 illustrates the molecular structure of sphingomyelin, which contains oleic acid and choline.

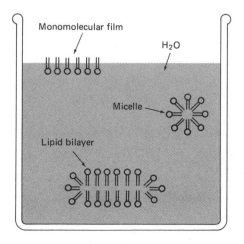

Fig. 7-8. A glycolipid.

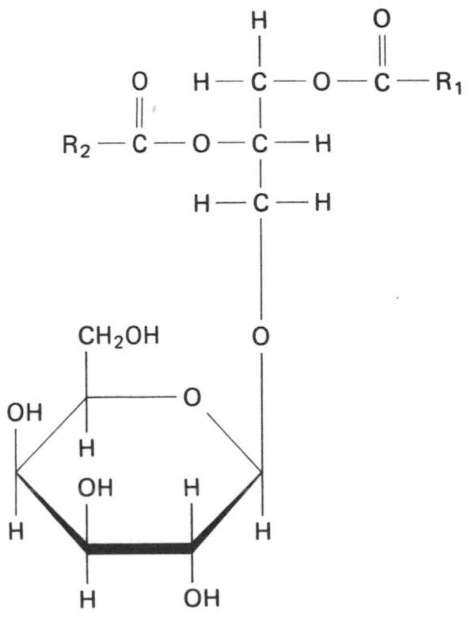

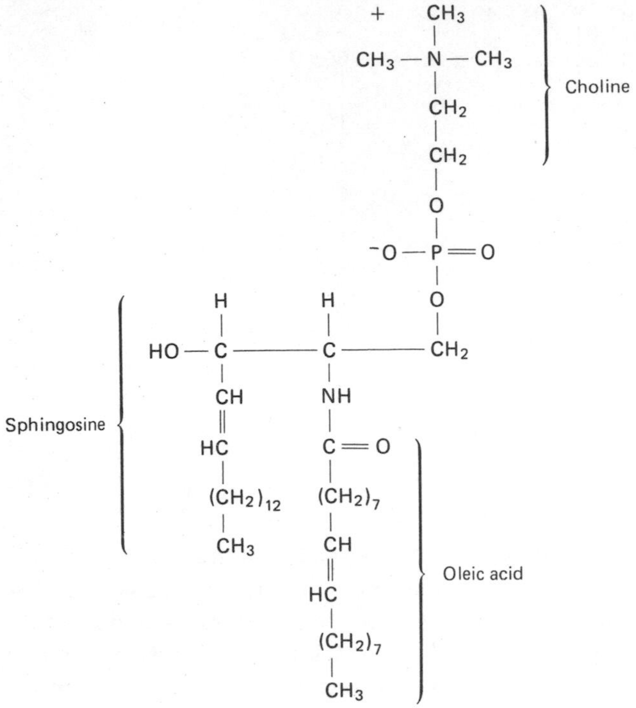

Fig. 7–9. Sphingomyelin, a sphingolipid.

Waxes

Waxes are synthesized by living cells and are usually excreted externally to serve as protective barriers against water loss and abrasion. All of the external surfaces of plant structures are covered with waxy substances, which act to minimize water loss by evaporation. Waxes are esters of long-chain fatty acids with long-chain primary alcohols. They usually exist in nature as complicated mixtures with long-chain hydrocarbons, free alcohols, and fatty acids. Their insolubility in water and the absence of double bonds make them chemically inert and excellent barriers to chemical attack or water loss.

Terpenoids

The terpenoids are the only nonsaponifiable lipids and include the terpenes and the steroids. What these molecules share in common is a five-carbon precursor molecule called an isoprenoid unit. Both terpenes and steroids are synthesized by the repeated condensation of isoprenoid units (Figure 7-10).

Beta carotene, an accessory photosynthetic pigment of green plants, is a terpene synthesized from eight isoprenoid units (Figure 7-11).

Cholesterol, bile acids, and testosterone are only a few of the many steroid compounds synthesized in this same basic fashion. Rings A, B, C, and D of cholesterol indicate the basic ring structure common to all steroids (Figure 7-12).

Fig. 7-10. The condensation of isoprenoid units to form long-chain terpenes and steroids.

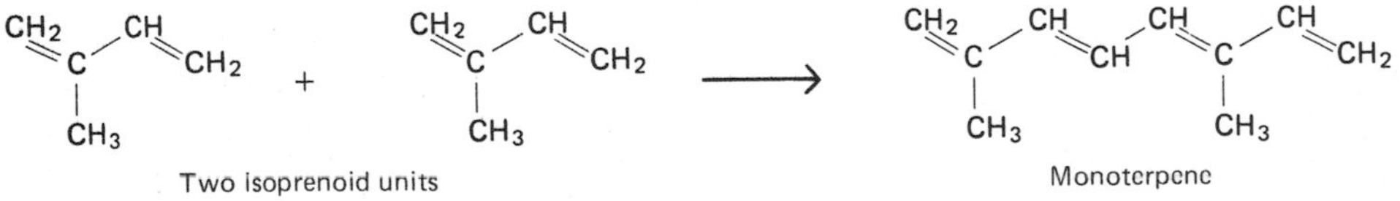

Two isoprenoid units   Monoterpene

Fig. 7-11. Beta-carotene, a 40-carbon terpene synthesized from eight isoprenoid units.

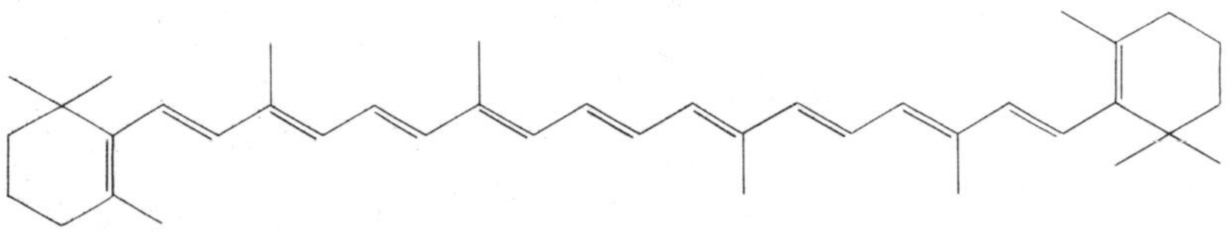

Lipid Analysis

Initially, progress in lipid research was not as rapid as that in other areas, such as proteins and carbohydrates, because of the special problems associated with the separation and analysis of water-insoluble substances. This, however, is no longer the case, thanks to the development of gas-liquid chromatography, lipophilic gel chromatography, and thin layer chromatography. A modification of thin layer chromatography called argentation chromatography is of special interest because it is quick and easy and achieves excellent separation of fatty acids and triglycerides. Silica gel impregnated with silver nitrate is used in this technique. As the solvent migrates up the TLC plate, the silver ions form coordination complexes with the unsaturated regions of the fatty acids and retard their movement. The net result is a separation based upon degree of saturation, the saturated fatty acids moving faster than the unsaturated fatty acids. By the same token, triglycerides with a high proportion of saturated fatty acids (SSS, SSU*) move more rapidly than their counterparts with more unsaturated fatty acids (UUS, UUU). Since all naturally occuring fats and oils contain mixtures of triglycerides composed of different fatty acids, argentation TLC is an excellent method for both qualitative and quantitative analysis of these lipids.

The exercises described below call for the analysis of several easily available lipids: lard, soybean oil, butter, and margarine. It is recommended that the exercises be completed during two laboratory periods in the following sequence.

Day 1. Saponification—completed to the point of adding soap cake to water and dissolving it. Solubility of lipids—this can be completed during the 1-hour hydrolysis step of saponification.

Day 2. Saponification—finish this part of the exercise by testing the properties of the soap solution. Fatty acid isolation Thin layer chromatography

---

*Abbreviations indicate the proportion of saturated and unsaturated fatty acids in a triglyceride, e.g., SSU = triglyceride with two saturated and one unsaturated fatty acids.

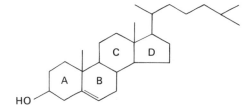

Fig. 7-12. Cholesterol, a typical steriod.

CAUTION: No flames in the laboratory—since several highly volatile and explosive solvents are used in these exercises, there should be no open flames in the laboratory at any time.

Lipids I. Saponification

Introduction
Your task is to prepare soap (fatty acid salts) from a plant or animal fat. Half of the laboratory groups will use lard (fat derived from pigs), and half will use soybean oil. The procedure in each case is the same, although the properties of the products will differ slightly. Saponification requires the refluxing of a lipid in the presence of alcoholic sodium hydroxide followed by evaporation of the alcohol. The resultant soap cake is dissolved in water to yield a soap solution. Once the reflux apparatus has been set up and the saponification is in progress, you should go on to part II.

Procedure
1. Transfer 10 g of lard (or soybean oil) to a 250-ml erlenmeyer flask containing 50 ml of 95% ethyl alcohol and 7 ml of 50% NaOH (w:v).
2. Seat a one-hole rubber stopper fitted with a length of glass tubing in the mouth of the flask. The glass tubing serves as a condenser during the boiling process.

CAUTION: Since the mixture is strongly basic, you should wear protective goggles throughout this entire exercise.

3. Lower the flask into a boiling water bath and clamp both the flask and the condenser to a ring stand (Figure 7-13).
4. Heat the mixture at 100°C for 60 minutes.

5. Remove the rubber stopper and condenser, allowing the mixture to continue boiling until all of the ethanol has evaporated. The presence of ethanol can be determined by gently wafting the vapors toward your nose and checking for a strong alcoholic aroma. *Do not in any case* put your nose directly over the flask while it is still boiling.

6. Cool the soap to room temperature, at which point it takes on the form of a semisolid cake in the bottom of the flask.

7. Break the cake into small pieces with a spatula and transfer the pieces to a 1000-ml beaker containing 500 ml of distilled water. Stir or gently shake the liquid until all of the soap is in solution. Since this may take several hours, it is advisable to use a magnetic stirrer or automatic shaker. Gentle warming of the liquid may also speed the process of dissolution.

Fig. 7-13. Laboratory setup for the saponification of a lipid.

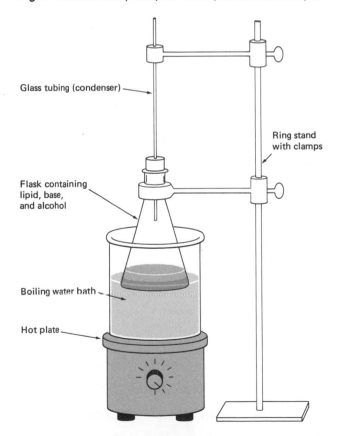

Glass tubing (condenser)

Ring stand with clamps

Flask containing lipid, base, and alcohol

Boiling water bath

Hot plate

8. Properties of the soap solution
    a. Check the pH of the solution with pH paper.
    b. Dip your fingers into the soap and rub them together. How does it feel? Rinse your fingers thoroughly with cold tap water.
    c. Transfer a few drops of the soap solution to a test tube containing 5.0 ml of distilled water. Shake the mixture vigorously and record your results.
    d. Place 5 drops of soybean oil in the bottom of a test tube, add 5 ml of distilled water, shake vigorously and record the result. Add 5 drops of soap solution, shake again and record your results.

Calculations and Questions

1. Have the free fatty acids neutralized all the NaOH in the mixture? On what do you base your answer?
2. In what ways does the soap mixture have the appearance and properties we would ordinarily associate with soaps? How do you explain these properties?

Lipids II. Solubility of Lipids

Introduction

Your job is to test the solubility of four different lipids (lard, soybean oil, margarine, and butter) in four different solvents. If the lipid does not go into solution after 5 minutes at room temperature, warm the mixture to determine whether that has any effect. Record all of your data in tabular form.

Procedure

1. Place a small amount (0.2 to 0.5 g) of lard in the bottom of four clean, dry test tubes.
2. Add about 4 ml of distilled water to the first tube and, by swirling and agitating the water, attempt to dissolve the fat. If solution has not occurred after 5 minutes at room temperature, report that result and then transfer the tube to a beaker containing hot (40 to 50°C) water. Continue your efforts to dissolve the fat at this temperature for 5 minutes and record your results.

3. Use this same procedure to determine the solubility of lard in chloroform, diethyl ether, and ethyl alcohol.
4. Repeat this procedure for each of the other lipids and report your data in tabular form.

Calculations and Questions
1. Draw the atomic structures of water, chloroform, diethyl ether, and ethyl alcohol. Explain your experimental results based upon the structures of these solvents.
2. What would you expect the solubility of fats and oils to be in such solvents as carbon tetrachloride, benzene, and methyl alcohol? Explain your answer.

Lipids III. Fatty Acid Isolation

Introduction
By means of acidification, you will precipitate the free fatty acids of the fatty acid soap prepared in the previous exercise. Since the fatty acids are insoluble and less dense than water, they aggregate on the surface of the solution, where they can be easily harvested and further purified. At the end of the isolation procedure, you should give half of your yield to one of the other laboratory groups so that each has a sample of lard fatty acids and a sample of soybean fatty acids.

Procedure
1. Stir the soap solution gently and add 1.0-ml aliquots of concentrated HCl, checking the pH after each addition, until the solution is very acidic (pH 2 to 3). In most cases this requires no more than 8 to 10 ml of HCl. When precipitation is complete, the fatty acids rise to the surface and aggregate in large, fluffy lumps.
2. Place a thermometer in the liquid, transfer the beaker to a hot plate, and heat (80 to 90°C) until the fatty acids are melted to an oily layer on top of the solution.
   a. Record the approximate temperature at which the majority of the fatty acids begins to melt.
   b. Notice the odor of volatile fatty acids which boil off at these temperatures.

3. Transfer the beaker to an ice bath and cool the liquid until the lipid layer congeals.
4. Skim the solidified fatty acid material off the top with a spatula and place it in a beaker containing 100 ml of distilled water.
5. Heat again on the hot plate, mixing the lipid with water to dissolve impurities, and then let the melted material rise and form a clear, oily layer on top of the water.
6. Cool in an ice bath as before and harvest the solidified fatty acid fraction.
7. Spread the fatty acid material on filter paper to dry and determine the total yield in grams.
8. Divide your sample in half and trade with another laboratory group so that each pair of students will have a sample of lard fatty acids and one of soybean fatty acids.

Calculations and Questions
1. Describe the properties of your fatty acid sample:
   a. What is its approximate melting point?
   b. Is it solid or liquid at room temperature?
   c. Is its solubility in polar and nonpolar solvents different from that of the parent fat or oil?
   d. Share these observations with the laboratory group to whom you have donated half of your sample.
2. What was your total yield of fatty acids and what percentage was that of the starting material?
3. Do you think that the fatty acids you have isolated represent all of the fatty acids present in the original lard or soybean oil sample? Explain your answer.
4. If you wanted to verify the presence of glycerol in the reaction mixture after saponification and acidification, in what fraction would you look for it?
5. If you were interested in isolating the fatty acids from a bar of handsoap, how might you go about it?

Lipids IV. Separation of Fatty Acids and Triglycerides by Argentation TLC

Introduction
The object of this exercise is to analyze the fatty acids and triglycerides of soybean oil and lard by chromatog-

raphy on silica gel TLC plates impregnated with silver nitrate. The number, position, and color of the separated lipids are used to qualitatively evaluate the differences and similarities of these lipids.

Procedure

1. Preparation of lipid samples (2% in petroleum ether)
   a. Dissolve 0.2 g of each lipid in 10 ml of petroleum ether (B.P. 35 to 60°C). The resulting four samples are:
      1) Lard fatty acids dissolved in petroleum ether
      2) Soybean oil fatty acids dissolved in petroleum ether
      3) Lard dissolved in petroleum ether
      4) Soybean oil dissolved in petroleum ether.
   b. Store the lipid solutions in sealed brown bottles.
2. Preparation of TLC plates
   a. Impregnation and activation of silica gel
      1) Obtain two TLC plates coated with silica gel (Baker-Flex Silica Gel IB2-F, 20 × 5 cm are adequate, but glass-backed plates would probably be superior).
      2) Pour a small amount of the silver nitrate solution (5% $AgNO_3$ in 50% methanol) into a shallow glass tray. Briefly dip each plate (silica side down) into the solution and when it is thoroughly wetted transfer it (face down) to a piece of clean paper towelling. Blot the excess liquid from both sides.
      3) Place the plates, silica side up, in an oven set at 100°C and bake for 60 minutes. When the plates have been activated (thoroughly dried), remove them from the oven and turn the oven up to 180°C for later use.
      4) Return unused silver nitrate solution to the bottle.
   b. Application of the samples
      1) Using a pencil, draw the origin and four points of sample application on each TLC plate in the same fashion as described in Chapter 6.
      2) On plate 1, apply two 5-μl samples of lard fatty acids to run parallel with two 5-μl samples of soybean oil fatty acids. Apply two samples (5 μl) of the lard solution to plate 2, running parallel with two samples (5 μl) of

the soybean oil solution. Observe all of the precautions previously described for the application of samples to silica gel plates (Chapter 6).
3. Chromatography

NOTE: All of the following operations should be performed in the hood.

   a. Prepare the solvent directly in the chromatography jar by adding 90 ml of petroleum ether (B.P. 35 to 60°C), 10 ml of diethyl ether, and 1 ml of glacial acetic acid. The solvent should fill the bottom of the jar to a depth of about 1 cm.
   b. Line the walls of the chromatography jar with a layer of filter paper and splash the solvent onto the paper, creating a solvent-saturated atmosphere inside the jar.
   c. Place both of the TLC plates in the chromatography jar, close the lid, and let the solvent rise to a height of 12 to 13 cm (1 hour).
   d. Remove the chromatograms, mark the solvent front, and hang them in the hood to dry for about 15 minutes.
4. Localization of lipid fractions by charring with sulfuric acid

CAUTION: This operation should be done with extreme caution because the locating solution is a 50 % solution of sulfuric acid, which will cause burns. After the plates have been dipped in the acid solution, do not handle them except with forceps.

   a. Pour a small amount of the acid solution (50% H_2SO_4, v:v) into a shallow glass tray. Dip the TLC plates in this solution and blot just as you did with the silver nitrate solution. Use forceps!
   b. Place the blotted plates (gel side up) on an enamel or glass tray and place them in a 180°C oven. Bake the plates for no more than 15 minutes. The combined effects of heat and acid will char all of the organic substances, which will now stand out as black spots.
   c. Immediately after the plates have cooled, circle all of the spots and measure the distances necessary to calculate R_f values. Continue to handle the plates with care because they contain resid-

ual sulfuric acid and are quite brittle as a result of this treatment. Include a drawing of both TLC plates in your notebook to show the pattern of lipid fractions.

Calculations and Questions

1. How many fatty acids are obvious in the lard and soybean fatty acid samples (TLC plate 1)?
2. Based upon the R_f values of the fatty acids, would you suggest that the two samples had any fatty acids in common? Are there fatty acids that appear to be unique to one or the other of the samples? Use drawings of the TLC plates to clarify your answers.
3. Bearing in mind that silver ions in the silica gel have the effect of retarding upward mobility of the unsaturated fatty acids, what might you guess to be the identity (or relative degree of saturation) of the fatty acids in lard and soybean oil? How could you verify the identity of the fatty acids experimentally?
4. How many different triglycerides are obvious in each of the samples (TLC plate 2)?
5. How do you explain the fact that in both soybean oil and lard there are more triglycerides evident than fatty acids?
6. Which triglycerides are common to the two samples and which are unique? Use a drawing of the TLC plate to explain your answer.
7. Do your results bear out the generalization that plant oils contain more unsaturated fatty acids than animal fats?

Additional Reading

F. B. Armstrong, *Biochemistry,* 2nd ed., Chaps. 12 and 19. Oxford University Press, New York, 1983.

C. B. Barrett, M. S. J. Dallas, and F. B. Padley, "The Quantitative Analysis of Triglyceride Mixtures by Thin Layer Chromatography on Silica Gel Impregnated with Silver Nitrate," *J. Amer. Oil Chem. Soc.* **40**:580 (1963).

E. E. Conn and P. K. Stumpf, *Outlines of Biochemistry,* 4th ed. Wiley, New York, 1976.

L. B. Dotti and J. B. Orten, *Laboratory Instructions in Biochemistry,* 8th ed. Mosby, St. Louis, 1971.

M. I. Gurr and A. T. James, *Lipid Biochemistry: An Introduction,* 2nd ed. Wiley, New York, 1975.

A. L. Lehninger, *Biochemistry,* 2nd ed. Worth, New York, 1975.

L. J. Morris, "Specific Separations by Chromatography on Impregnated Absorbants," in *New Biochemical Separations,* A. T. James and L. J. Morris, eds., Chap. 14. Van Nostrand, New York, 1964.

G. H. Pritham, *Anderson's Laboratory Experiments in Biochemistry.* Mosby, St. Louis, 1968.

R. W. Schery, *Plants for Man,* 2nd ed. Prentice-Hall, Englewood Cliffs, NJ, 1972.

V. P. Skipski and M. Barclay, "Thin Layer Chromatography of Lipids," in *Methods in Enzymology,* J. M. Lowenstein, ed., Vol. 14. Academic Press, New York, 1969.

8 NUCLEIC ACIDS: ISOLATION AND CHARACTERIZATION OF *E. COLI* DNA

Nucleic Acids

Deoxyribonucleic acid (DNA) and ribonucleic acid (RNA) are the informational molecules of the cell which specify and direct the synthesis of proteins. They are high-molecular-weight polymers of nucleotides and resemble one another in general makeup but have significant structural and functional differences. DNA is a double-stranded molecule found in the nucleus of eucaryotic cells and functions as the genetic material of the cell. RNA, a single-stranded molecule which directs protein synthesis, is found predominantly in the cytoplasm.

The informational content of DNA is in the form of nucleotide sequences which determine the sequence of amino acids in proteins. The DNA nucleotide sequence is converted into an amino acid sequence in a two-phase process. The first stage is the synthesis of an RNA molecule, using the DNA sequence as a template or model. This enzymatically mediated step occurs in the nucleus and is called transcription. After the RNA has been transported to the cytoplasm, the second stage, which is called translation, begins. Translation is the precise assembly of proteins from amino acids and is mediated by several types of enzymes, RNAs, and other cofactors. In summary, the process involves the synthesis of RNA (the nucleotide sequence of which is specified by the DNA sequence) and the manufacture of protein (the amino acid sequence of which is determined by the RNA sequence). The informational flow has come to be known as the Central Dogma of molecular biology (Figure 8–1). Although Figure 8-1 does not indicate as much, informational flow has been demonstrated to occur in other directions, including DNA → DNA (DNA replication) and RNA → DNA (reverse transcription).

Nucleotides

The building-block molecule of all nucleic acids is the nucleotide (Figure 8-2), which contains a phosphorylated five-carbon sugar (ribose or 2-deoxyribose) linked to a nitrogen-containing base (purine or pyrimidine). The molecule illustrated in Figure 8-2 contains ribose phosphorylated in the 5' position and the nitrogenous base guanine (a purine). This nucleotide is called guanosine 5'-monophosphate or, more simply, GMP. Other nucleotides found in DNA and RNA vary with respect to the sugar (ribose or 2-deoxyribose), the nitrogenous base (adenine, guanine, cytosine, thymine, or uracil), and the position of the phosphate on the sugar (2', 3', or 5').

Nucleic Acid Structure

In DNA and RNA the nucleotides are joined by phosphodiester bonds between the 5' position of one sugar and the 3' position of the next (Figure 8-3). The result is a linear molecule consisting of a sugar-phosphate

Fig. 8-1. The informational flow in a cell, which illustrates how DNA directs the synthesis of proteins.

$$\text{DNA} \xrightarrow{\text{transcription}} \text{RNA} \xrightarrow{\text{translation}} \text{PROTEIN}$$

backbone with purines and pyrimidines extending outward from the sugar residues. Figure 8-3 represents a short stretch of RNA, which contains the sugar ribose and the bases adenine, guanine, cytosine, and uracil. Other bases are found in RNA, but these four predominate.

DNA in its native configuration consists of two polynucleotide strands regularly twined around one another and around a central axis to form a double helix. The two strands are held together by hydrogen bonds between the nitrogenous bases of the two molecules, as shown in Figure 8-4. Notice from this figure that the two nucleotide chains run in opposite (antiparallel) directions and that the only base pairs sterically consistent with a regular double helix are adenine-thymine and guanine-cytosine. The only sugar found in DNA is 2-deoxyribose. As indicated in Figure 8-5, the bases which predominate in DNA are adenine (A), guanine (G), cytosine (C), and thymine (T). In this illustration the sugar-phosphate backbones are represented by helical ribbons and the bases are sticklike projections extending into the core of the double helix. Since the only base pairs allowed are A:T and G:C, the sequence of nucleotides on one strand is the complement of the sequence on the other, and one can easily infer the order of bases on one strand from the other.

DNA has been isolated from an extremely wide variety of organisms, and although the general features of the molecule herein described are common to all living things, the critical way in which the genetic material differs is in the order of bases on the polynucleotide chain. The genetic uniqueness of DNA isolated from various species lies in the different base sequences which code for the manufacture of different proteins. As previously discussed, the proteins are the "workhorses" of the cell, and their proper functioning is intimately connected with their primary structure. If the DNA base sequence is correct and if transcription and translation proceed normally, the protein will be assembled correctly and function normally. On the other hand, if there are changes or errors in either the base sequence or protein synthesis, the newly formed protein will have an incorrect amino acid sequence and in all probability will not function properly. It seems quite clear then that a detailed understanding of the structure and functioning of nucleic acids is central to an understanding of life at the molecular level.

Physical and Chemical Properties of DNA

In its native, double-stranded state, DNA is a high-molecular-weight (10^6 to 10^9 daltons) polymer which may exist either as a circle or as a long, slender rod. Owing to the rigidity of these molecules and their enormous length-to-width ratio, solutions of DNA are quite viscous and the molecules are subject to degradation by

Fig. 8-2. Guanosine-5′-monophosphate, a nucleotide found in both DNA and RNA.

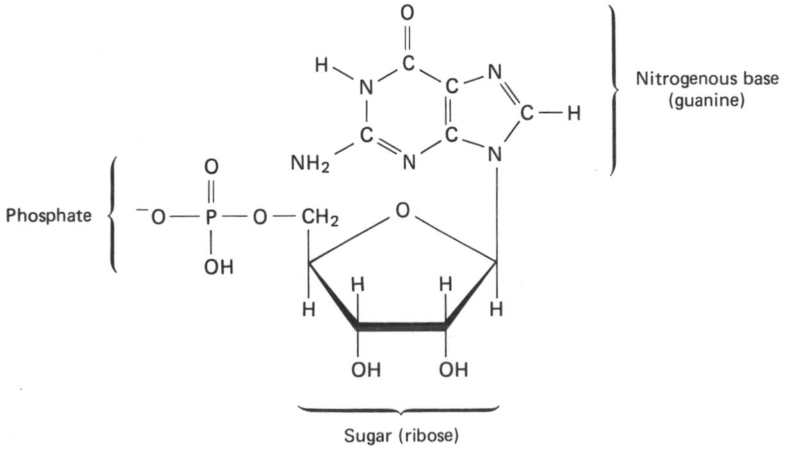

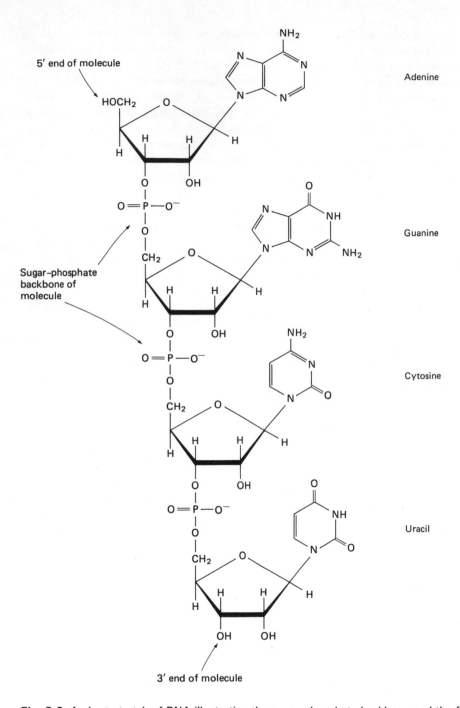

Fig. 8-3. A short stretch of RNA illustrating the sugar-phosphate backbone and the four bases commonly found in RNA.

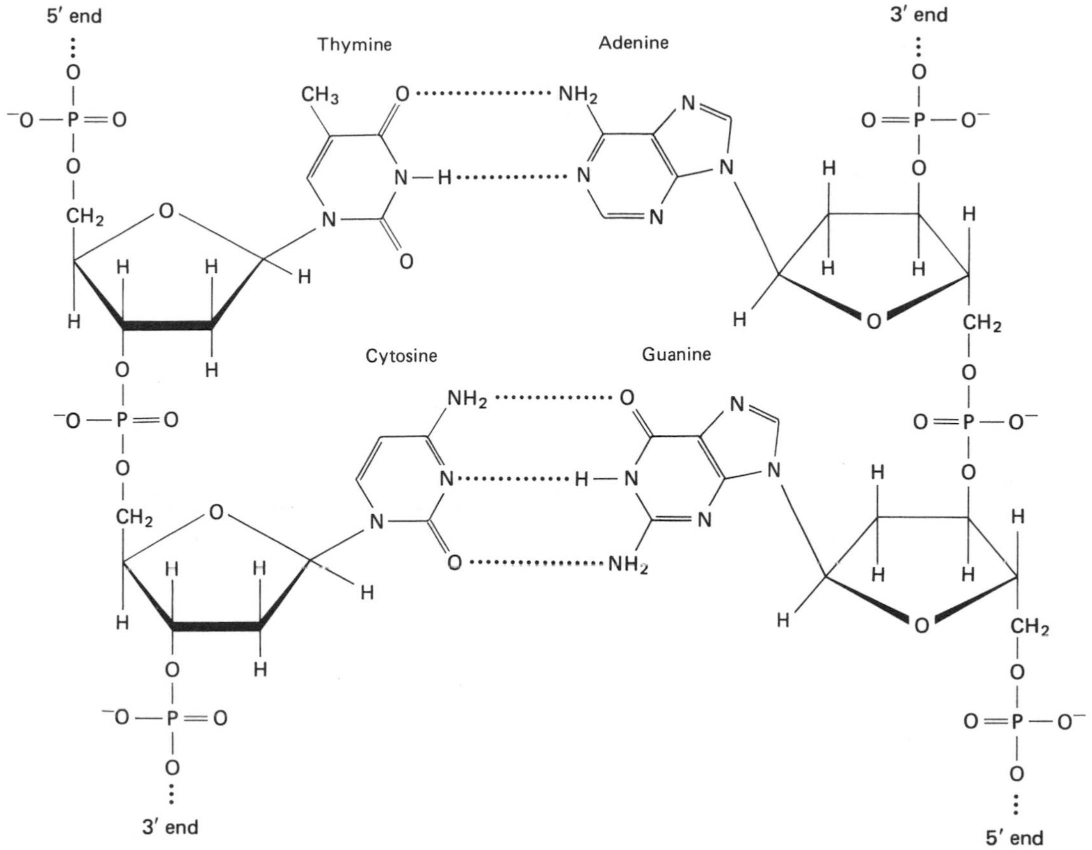

Fig. 8-4. Two DNA strands, running from the top of figure to the bottom but in opposite directions. The dotted lines indicate the hydrogen bonds between base pairs of the two strands.

shear forces. Consequently, a great deal of care must be exercised in the handling of DNA solutions because with each transfer by pipette there is an excellent chance of shearing the DNA and reducing its molecular weight. As might be expected, the gentlest extraction procedures yield DNA most nearly native in terms of molecular size.

Since the hydrophilic sugar-phosphate backbones of the DNA molecule are on the outside of the double helix and the hydrophobic bases extend inward, DNA tends to be fairly soluble in water. However, at high concentrations DNA exists as a colloidal suspension in aqueous media.

As a result of the presence of the nitrogenous bases, which absorb light strongly in the ultraviolet range (250 to 275 nm), the nucleic acids have a peak absorbance at 260 nm (Figure 8-6). This fact has been usefully exploited in the development of numerous qualitative and quantitative tests for DNA and RNA.

Under normal conditions the DNA molecule is quite stable, but at high temperatures (80 to 100°C) and pH values outside the range of 4 to 10, the hydrogen bonds break and the two polynucleotide strands separate, or denature (Figure 8-7). Denatured DNA is single-stranded, and its properties are markedly different from those of native DNA. Denaturation is most notably marked by an abrupt increase in UV absorption (hyperchromicity) and a decrease in viscosity. Native DNA typically exhibits a hyperchromic shift of 30 to 40 per cent when denatured (Figure 8-8). This is often used as

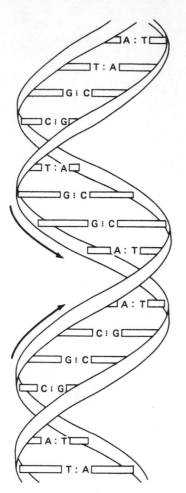

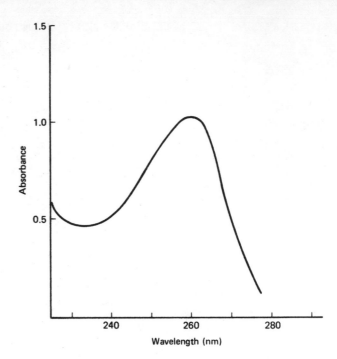

Fig. 8-6. The absorption spectrum for native DNA.

Fig. 8-5. Two antiparallel strands of DNA showing the double helix and the base pairing that holds the two molecules together.

Fig. 8-7. The progress of heat denaturation of DNA.

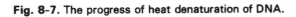

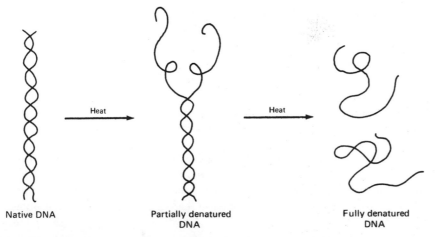

Native DNA

Partially denatured
DNA

Fully denatured
DNA

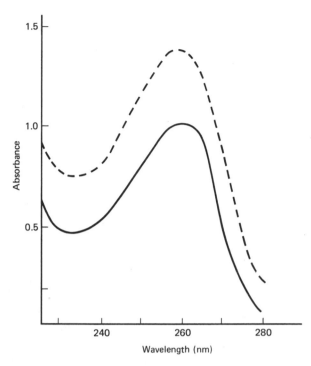

Fig. 8-8. The absorption spectra for native DNA (——) and denatured DNA (– – – –).

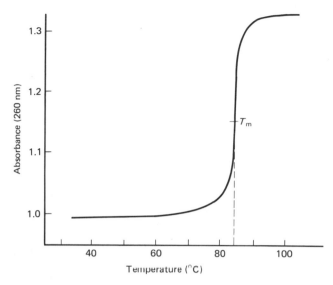

Fig. 8-9. The melting profile for a sample of native DNA. The melting temperature (T_m) for this sample is 84°C.

a measure of purity or double-strandedness of DNA samples because those containing single-stranded RNA or denatured DNA exhibit lower hyperchromic shifts.

Heat denaturation of DNA in a UV spectrophotometer equipped with a thermostatically controlled sample compartment provides a method for monitoring strand separation. The DNA is gradually heated from room temperature to 100°C while a continuous record of the UV absorbance is made. Interestingly enough, the hyperchromic shift does not occur gradually; rather it happens suddenly across a narrow temperature range (Figure 8-9), the median of which is called the melting temperature (T_m). Melting temperatures for DNA isolated from many different organisms have revealed that, while the T_m for a specific DNA is constant, there is much variation from species to species. The variance in T_m is a reflection of the different base compositions of the DNAs. Those DNAs with a high percentage of guanine and cytosine exhibit higher melting points than those with a high percentage of adenine and thymine,

which mirrors the greater stability of the guanine-cytosine pairs.

Alkaline conditions (pH > 10) also cause strand separation and the characteristic increase in UV absorption. An interesting difference between DNA and RNA lies in their reaction to prolonged alkaline conditions. DNA maintained in an alkaline solution overnight at 37°C exhibits little or no change other than the expected denaturation because the phosphodiester linkages of DNA are resistant to this treatment. RNA kept under the same conditions is hydrolysed to low-molecular-weight mononucleotides. The difference in sensitivity to alkaline conditions is often used to separate DNA from RNA.

Nucleic Acids I. Isolation of DNA from *E. coli* Cells

Introduction

The object of this experiment is to extract high-molecular-weight DNA from lyophilized (freeze-dried) *E. coli* cells, using an extraction procedure published by Julius Marmur in 1961. It is your task to read the article in its entirety and to develop your own laboratory protocol for the extraction of DNA. After your plan has been approved by the laboratory instructor, you will be provided with a sample of 2 to 3 g of wet-packed *E. coli* cells and the appropriate solutions and equipment and then expected to isolate DNA from the cells. The reason for this departure from the method of previous chapters is to expose you to a more realistic laboratory situation. A biochemist confronted with the problem of designing new experimental procedures cannot depend upon detailed laboratory manuals but must go to the literature and develop techniques based upon the published work of other investigators.

Marmur's article was chosen for several reasons. It is an unusually detailed account and includes methods, precautions, and alternative methods as well as a description of the role each reagent plays in the procedure. It addresses the problem of extracting bacterial DNA from many different species and includes suggestions for the isolation of DNA from both higher animals and plants. It is truly a landmark publication because it describes a highly adaptable procedure for the extraction of DNA from living things. A quick perusal of the DNA literature since 1961 will immediately testify to the impact of this work.

Procedure

Read through the entire Marmur article once and then go back and read the Materials and Methods section in detail. As you read this part of the paper, answer the series of questions posed below. They are designed to guide your reading and understanding of the article. When you are finished, write up a step-by-step scheme of what you plan to do in the laboratory to extract DNA from *E. coli* cells. It may be helpful if you reexamine the procedure and flow sheet for the isolation of protein described in Chapter 4. Once your protocol has been approved by the laboratory instructor, you may proceed with the isolation.

The following questions can all be answered by a careful reading of the Marmur paper.

1. Describe the role of each of the following reagents used in the isolation of DNA.
   a. EDTA
   b. Sodium lauryl sulfate
   c. Saline citrate
   d. Ethyl alcohol
   e. Ribonuclease
2. At what temperature should the extraction procedure be conducted? Are there any exceptions?
3. Which method of cell lysis is the more appropriate for *E. coli* cells: detergent lysis alone (using sodium lauryl sulfate) or lysozyme digestion followed by detergent lysis?
4. If the final volume of the lysed cell suspension (after the heat step) is 30 ml, how much 5.0 *M* sodium perchlorate must be added to bring the final concentration to 1.0 *M* sodium perchlorate?
5. The heat treatment recommended in this procedure calls for 60°C for 10 minutes. Based upon the T_m of *E. coli* DNA, how much higher could you elevate the temperature for this step?
6. After the emulsion has been separated into three layers by centrifugation, which layer contains the DNA?
7. If no DNA "spools" out after the addition of ethanol, what should be done to collect the DNA?
8. What point in the extraction would be an appropriate place to interrupt the procedure?
9. How should the purified DNA be stored?
10. Do you think that the use of lyophilized (freeze-dried) *E. coli* cells will have any adverse effect on your results?
11. Based upon the length of your laboratory period, how many periods should be set aside for this exercise?

A Procedure for the Isolation of Deoxyribonucleic Acid
from Micro-organisms

J. MARMUR

Department of Chemistry, Harvard University
Cambridge, Massachusetts, U.S.A.
(Received 6 December 1960)

A method has been described for the isolation of DNA from micro-organisms which yields stable, biologically active, highly polymerized preparations relatively free from protein and RNA. Alternative methods of cell disruption and DNA isolation have been described and compared. DNA capable of transforming homologous strains has been used to test various steps in the procedure and preparations have been obtained possessing high specific activities. Representative samples have been characterized for their thermal stability and sedimentation behaviour.

1. Introduction

To facilitate the study of the biological, chemical and physical properties of DNA it is necessary to obtain the material in a native, highly polymerized state. Several procedures have described the isolation of DNA from selected groups of micro-organisms (Hotchkiss, 1957; Zamenhof, Reiner, DeGiovanni & Rich, 1956; Chargaff, 1955). However, no detailed account is available for the isolation of DNA from a diverse group of micro-organisms. The reason for this is that micro-organisms vary greatly in the ease with which their cell walls can be disrupted, in their content of capsular polysaccharides (which are difficult to separate from DNA) and in the association of DNA to protein which influences the ease of DNA purification (Kirby, 1957). Most of these difficulties have been overcome in the present procedure which has been applied successfully to approximately 50 different species of micro-organisms. Included in this number are those organisms whose DNA can transform homologous and closely related strains and that have thus provided a very useful tool in determining the efficacy of many of the steps outlined in the procedure.

In general, the method to be described can be outlined as follows: the cells are first disrupted, the cell debris and protein removed by denaturation and centrifugation, the RNA removed by RNase and the selective precipitation of the DNA with *iso*-propanol. Degradation by DNase and divalent metal ion contamination is prevented by the presence of chelating agents and by the action of sodium lauryl sulfate. The products obtained, although they may not reflect the *in vivo* molecular weight, have molecular weights in excess of 6×10^6 and a high specific transforming activity, where this is present.

2. Materials and Methods
(a) Reagents
Saline-EDTA, 0·15 M-NaCl plus 0·1 M-ethylenediaminetetra acetate (EDTA), pH 8. The EDTA, and/or high pH, inhibit DNase activity.

This work was supported by a grant from the U.S. Public Health Service (C-2170).

Author's present address is Department of Biochemistry, Albert Einstein College of Medicine, Bronx, N.Y.

Sodium lauryl sulfate, 25%. The anionic detergent ($NaC_{12}H_{26}SO_4$) will lyse most nonmetabolizing cells, inhibit enzyme action and denature some proteins (Bayliss, 1937; Bolle & Kellenberger, 1958).

Lysozyme, crystalline (Armour). Used to lyse cells resistant to detergent action. Cells lysed with lysozyme are then subjected to sodium lauryl sulfate as well.

Sodium perchlorate, 5 M. The high salt concentration provided by the perchlorate (Lerman & Tolmach, 1957) helps dissociate protein from nucleic acid.

Chloroform-isoamyl alcohol, 24:1 (v/v). Used to deproteinize, according to the method of Sevag, Lackmann & Smolens (1938). The chloroform causes surface denaturation of proteins. The *iso*amyl alcohol reduces foaming, aids the separation, and maintains the stability, of the layers of the centrifuged, deproteinized solution.

Ethyl alcohol, 95%. Used to precipitate nucleic acids following deproteinization. Denatured ethyl alcohol may also be used.

Saline-citrate, 0·15 M-NaCl plus 0·015 M-trisodium citrate, pH 7·0 $\pm$ 0·2. Maintains ionic strength of dissolved DNA and chelates divalent ions.

Dilute saline-citrate, 0·015 M-NaCl plus 0·0015 M-trisodium citrate. DNA dissolves more readily in dilute salt solutions but should *never* be dissolved in pure water.

Concentrated saline-citrate, 1·5 M-NaCl plus 0·15 M-trisodium citrate. The concentrated solution is used to bring the dilute saline-citrate solute, in which the nucleic acid is dissolved, up to saline-citrate concentration. The volume added need only be approximate until the final pure product is obtained.

Ribonuclease, 0·2% (crystalline, Armour) in 0·15 M-NaCl, pH 5·0. The solution is heated at 80°C for 10 min to inactivate any contaminating DNase. The RNase digests the RNA and facilitates its separation from DNA.

Acetate-EDTA, 3·0 M-sodium acetate plus 0·001 M-EDTA, pH 7·0. This provides the proper ionic environment in the *iso*propanol step for the separation of DNA from RNA or its digestion products (Simmons, personal communication).

Isopropanol. Used to precipitate DNA selectively; RNA remains in solution. In some cases it will selectively precipitate and separate DNA from polysaccharides.

(b) Equipment for DNA Isolation

Centrifuges, Servall SS-1 operating at 5 to 10,000 rev/min (3,000 to 13,000 g) and a clinical swinging bucket centrifuge capable of spinning at 2 to 3000 rev/min (300 to 600 g).

Glass-stoppered flasks, for deproteinization.

Shaker, wrist action or reciprocal for deproteinization. Several hundred strokes/min.

Volumetric pipette, 10 to 15 ml. fitted with an 18 in. (approx.) rubber tube attached to the upper end. Used to remove the aqueous layer from the deproteinized, centrifuged mixture.

Stirring motor, fitted with a glass stirring rod with a screw taper. Used to stir the solution (500 to 1,000 rev/min) during the *iso*propanol addition.

(c) Physical and Biological Measurements

Determination of T_m. The method has been previously described (Doty, Boedtker, Fresco, Haselkorn & Litt, 1959; Marmur & Doty, 1959).

Determination of sedimentation coefficient and molecular weight. The sedimentation coefficient, $S_{20,w}$, of DNA dissolved in standard saline-citrate was determined in the Spinco ultracentrifuge model E at a concentration of 20 μg/ml. at a speed of 35,600 rev/min using ultraviolet optics. The centrifuge cell was fitted with a Kel-F centerpiece. The molecular weight of the sample can then be obtained using the relationship established by Doty, McGill & Rice (1958):

$$S_{20,w} = 0 \cdot 063 \, M_w^{0.37}$$

Transformation. The transformation of *Diplococcus pneumoniae* was carried out by the method of Fox & Hotchkiss (1957) using transformable, glycerol-treated, cells stored at -20°C. When it was found necessary to examine the biological properties of *Bacillus subtilis* DNA, the method of Spizizen (1959) was used to transform this organism.

(d) Isolation Procedure

The procedure is designed for 2 to 3 g wet packed cells. The volumes are only approximate unless otherwise stated. All operations can be performed at room temperature except the RNase treatment, which is carried out at 37°C.

Bacteria grown to the logarithmic phase of their growth cycle are harvested by centrifugation and washed once with 50 ml. saline-EDTA. After collecting by centrifugation, the cells are suspended in a total volume of 25 ml. of saline-EDTA. Lysis[1,2,3†] is effected by the addition of 2·0 ml. sodium lauryl sulfate and the mixture placed in a 60°C[4] water bath for 10 min then cooled to room temperature. Lysis of the culture results in a dramatic increase in viscosity[5] accompanying the release of the nucleic acid components and some clearing. If, on preliminary testing the cells are insensitive to the detergent, but sensitive to lysozyme, approximately 10 mg of lysozyme are added to the cells suspended in saline-EDTA. The mixture is then incubated at 37°C with occasional shaking and the lysis followed by noting the increase in viscosity. In some cases 30 to 60 min may be required for optimum results. When lysozyme is used sodium lauryl sulfate is added as well, *after* the cells have lysed, followed by the 60°C heating and cooling.

Perchlorate is added to a final concentration of 1 M to the viscous, lysed suspension and the whole mixture shaken with an equal volume of chloroform-*iso*amyl alcohol in a groundglass stoppered flask for 30 min.[6] The resulting emulsion[7] is separated into 3 layers by a 5-min centrifugation at 5,000 to 10,000 rev/min in the Servall. The upper aqueous phase contains the nucleic acids and is carefully pipetted off into a tube or narrow flask. The nucleic acids are precipitated by gently layering approximately 2 vol ethyl alcohol on the aqueous phase. When these layers are gently mixed with a stirring rod, the nucleic acids "spool" on the rod as a threadlike precipitate[8] and are easily removed. The precipitate is drained free of excess alcohol by pressing the spooled rod against the vessel. The precipitate is then transferred to approximately 10 to 15 ml. of dilute saline-citrate[9]

†Numbers refer to the following section on Procedure notes.

and gently removed from the stirring rod by swirling it back and forth. The solution is gently shaken or pipetted until dispersion is complete (lumps can be recognized by adhering air bubbles when the solution is shaken). The solution is adjusted approximately to standard saline-citrate concentration by adding concentrated saline-citrate, shaken as before with an equal volume of chloroform-*iso*amyl alcohol for 15 min, centrifuged[10] and the supernatant removed. It is then deproteinized repeatedly with chloroform-*iso* amyl alcohol,[11] as described, until very little protein is seen at the interface.

The supernatant obtained after the last in the series of deproteinizations, is precipitated with ethyl alcohol and dispersed in saline-citrate (about 0·5 to 0·75 the supernatant volume) in the manner already described. Ribonuclease[12] is added to a final concentration of 50 μg/ml. and the mixture incubated for 30 min at 37°C. Following the digestion of the RNA it becomes possible to remove protein which resisted earlier chloroform deproteinizations. The digest is again subjected to a series of deproteinizations until there is little or no denatured protein visible at the interface after centrifugation. The supernatant, after the last such treatment, is again precipitated with ethyl alcohol and the drained nucleic acid dissolved in 9·0 ml. dilute saline-citrate. When solution has occurred, 1·0 ml. acetate-EDTA is added and while the solution is rapidly stirred, 0·54 vol[13] *iso*propyl alcohol is added dropwise into the vortex. The DNA usually precipitates in a fibrous form after first going through a gel phase at about 0·5 vol *iso*propyl alcohol. RNA or oligoribonucleotides and cellular or capsular polysaccharides remain behind, while the DNA threads wind around the glass propeller. If the yield is good, the DNA is redissolved and precipitated once more with *iso*propanol in the manner described. The final precipitate is washed free of acetate and salt by gently stirring the adhered precipitate in progressively increasing (70 to 95%) portions of ethyl alcohol, and is then dissolved in the solvent of choice. If the solution is not clear, it can be clarified by centrifuging in the Servall centrifuge for 10 min at 5,000 rev/min.[14]

By using caution and recovery steps, up to 50% of the DNA from the cell is obtained; in general 1 to 2 mg of DNA is obtained from 1 g wet packed cells.[15] The DNA can be stored in solution at 5°C in the presence of several drops of chloroform.[16,17,18] If it is so desired, the DNA (free of spores) can be sterilized by exposure to 75% ethyl alcohol for several hours and then transferred to a sterile solvent.

Notes on the Procedure for the Isolation of DNA

It would be very difficult to describe a definitive technique for the efficient isolation of DNA from a wide variety of micro-organisms. The method described can undoubtedly be modified or improved to eliminate difficulties encountered with specific strains. In general, the Gram negative organisms yield themselves readily to the procedure resulting in good recoveries of highly polymerized DNA. Several suggestions are offered to improve yields and eliminate some of the difficulties that may arise.

1. A spot test on a centrifuged portion of the culture should be made to determine whether the organisms are susceptible to sodium lauryl sulfate, lysozyme or neither of the two. If lysed by both detergent and enzyme, the former is preferable since DNase is inactivated in its presence. If lysozyme is used, the detergent is added *after* maximum enzymatic lysis is attained.

2. Organisms that are readily lysed by sodium lauryl sulfate include:—all Gram negative strains thus far encountered (Enterobacteriaceae, *Hemophilus influenzae, Rhizobium Japonicum, Pseudomonas aeruginosa, Pasteurella pestis*) as well as *D. pneumoniae* (which also lyses readily with deoxycholate) *Bacillus stearothermophilus, B. macerans, B. brevis, B. licheniformis, Clostridium madisoni, Cl. Chauvei, Cl. butylicum, Rhodospirillum rubrum, Micrococcus lysodeikticus* and *Mycoplasma* (PPLO). *Euglena gracilis* and *Chlamydomonas reinhardii* and the slime mold *Dictyostelium discoideum* are readily lysed by the detergent. Some strains of *Streptococcus* and of *M. pyogenes* var *aureus* lyse slowly with sodium lauryl sulfate and cell disruption is sometimes facilitated by raising the temperature of the lysing mixtures to between 70° and 75°C. The former genus can also be lysed by extracts from *Streptomyces albus* (McCarty, 1952).

The organisms lysed with lysozyme include: *B.*

subtilis, B. natto, B. cereus, B. megaterium, Cl. perfringens as well as the Actinomycetes (*S. albus* and *S. viridochromogenes*) which lyse very slowly with the enzyme (Sohler, Romano & Nickerson, 1958), depending on the state of growth when harvested.

It has been found that some strains of *Streptococcus* and of *M. pyogenes* var *aureus, Lactobacillus acidophilus* and baker's yeast are insensitive to lysis by either lysozyme or detergent. Yeast cells can be lysed by an extract from the snail *Helix pomatia* (Eddy & Williamson, 1957). If no means of enzyme or detergent lysis is available, the cells can be disrupted by grinding with alumina or glass powder (see below). The product, however, usually has a lower molecular weight than that obtained by the other methods described. (For other means of cell wall disruption see review by Weibull, 1958.)

The method has been successfully employed in the isolation of DNA from animal tissue, using sodium lauryl sulfate to lyse the cells.

3. It has been noted that some cells grown in the presence of 5-bromodeoxyuridine have altered their susceptibility to lysis. Thus, *D. pneumoniae* grown in the presence of this analogue will lyse slowly if at all with sodium lauryl sulfate, but remains susceptible to deoxycholate. This situation has not arisen in the case of *E. coli*.

4. If potent nuclease action is anticipated (e.g.*Serratia marcescens,*) the detergent lysed suspension should be heated to within 10°C of the DNA T_m to eliminate its activity.

5. Some organisms (e.g. *Klebsiella pneumoniae*) are difficult to harvest and when lysed give rise to extremely viscous solutions. This is due to polysaccharide which is usually eliminated in the final stages of the DNA isolation but may also be removed earlier by initial use of the *iso*propanol step.

6. The shaking in the first deproteinization is carried out for 30 min because of the high viscosity of the lysate; subsequent steps are for 15 min.

7. A very critical step in the procedure is the concentration of cells being lysed. Too low a cell concentration will give rise to losses in subsequent alcohol precipitation whereas, if the cell suspension is too thick, the first chloroform deproteinization will result in a "lumpy" emulsion. The lumps consist of denatured protein with large amounts of occluded DNA

and form a voluminous middle layer when the mixture is centrifuged. This difficulty is easily remedied in subsequent preparations by lysing a more dilute suspension of cells. In order to recover DNA from the denatured protein layer, because of excessive occlusion or when a high yield of DNA is desired under normal circumstances, the residue is shaken with a 10 ml. portion of dilute saline citrate for 15 min, centrifuged and the precipitate obtained from the ethyl alcohol addition combined with the remainder of the preparation.

8. If difficulty is encountered in collecting a majority of the nucleic acid threads following ethyl alcohol addition or if no threads appear because of DNA degradation, the solution is subjected to a short centrifugation, the sediment dissolved in a small volume of dilute saline-citrate and the alcohol precipitate added to the remaining portion of the preparation.

9. When dissolving the fibrous nucleic acid precipitate it is well to keep the concentration of DNA at a level of about 0·2 to 0·8 mg per ml. Too low a concentration results in degradation (Hershey & Burgi, 1960) and loss of biological activity during handling. Higher concentrations are highly viscous and difficult to handle and disperse. In the early stages of the preparation, the redissolved precipitate gives rise to turbid suspensions; as the purification proceeds, the nucleic acid takes on a glassy appearance when being dispersed and results in clear solutions.

10. If the emulsion separates readily into two phases after shaking, centrifugation in a swinging bucket clinical centrifuge at 2 to 3000 rev/min is sufficient to clear the aqueous layer of chloroform and most of the denatured protein.

11. The best time to interrupt the procedure, if it should be necessary, is after any of the deproteinization steps. In this case store the uncentrifuged emulsion until the procedure can be resumed. The average length of time required to isolate purified DNA starting from the lysis of the cells is approximately 5 to 8 hr.

12. It is also possible to isolate RNA by omitting the RNase treatment, recovering the ethyl alcohol precipitates by centrifugation and saving the solution after the DNA is removed by the *iso*propanol step.

It would however be best to carry out the *iso*propanol step earlier in the procedure.

13. Several cases have been encountered where the *iso*propanol step does not precipitate the DNA when 0·54 vol have been added (*C. reinhardii* and in one case of DNA isolated from tobacco leaves). When this occurs, larger volumes of *iso*propanol should be added. At times the precipitate is granular and can be collected by centrifugation.

14. It has been repeatedly observed that the purified DNA isolated from some spore formers (e.g. *B. brevis, B. subtilis,* etc.) is contaminated with viable spores. The number of spores can be reduced or eliminated by harvesting the cells when spore formation is at a minimum, centrifuging the purified DNA for about 10 min at 10,000 rev/min in the Servall and/or treatment of the DNA with phenol (see 17).

15. The method can be applied to bacterial harvests of the order of 100 g but, to avoid the awkwardness of handling large volumes, the amount of reagents used can be scaled up less than proportionately.

16. If protein removal is incomplete, storage of the DNA solution in the presence of chloroform will sometimes leave a halo of denatured protein surrounding the interface of the solutions. The protein can be removed by centrifugation.

17. The introduction of the phenol method (Kirby, 1957) to isolate DNA prompted its use on purified DNA. Purified *D. pneumoniae* DNA shaken with water saturated phenol remained undegraded and did not show any apparent loss in biological activity. Trypsin and chymotrypsin likewise had little or no effect on the molecular weight or biological activity.

18. The amino acid content (other than glycine) of the purified DNA on acid hydrolysis is approximately 0·3 to 0·5% (determined by Dr. H. Van Vunakis on the Spackman, Stein & Moore (1958) automatic amino acid analyser). Typical ratios for absorption of the DNA at 260:230:280 mμ are 1·0:0·450:0·515.

(e) Alternative Methods of Cell Disruption and Lyophilization
Mechanical disruption

Various techniques have been described (Gunsalus, 1957) for the mechanical disruption of cells for the isolation of enzymes, nucleic acids, etc. An attempt was made to disrupt *D. Pneumoniae* cells by grinding with glass powder as well as by sonic treatment and to isolate the DNA from the disrupted cells according to the method described above. Since these and other methods might be applied to cells that resist enzyme and detergent action, we have applied them to *D. pneumoniae* as a basis of comparison to the detergent method of cell lysis.

Grinding with glass or alumina (particle size 500 mesh) is carried out at 5°C. The harvested, washed cells (with saline EDTA) are placed in a pre-cooled mortar, an equal weight of glass powder (Fisher Scientific Co.) added and the mixture ground with a pestle for 5 to 10 min. Ten vols of cold saline-EDTA, containing 2% sodium lauryl sulfate, are added and the suspension centrifuged to remove glass and large cell debris. The supernatant is then treated in the manner described above for the isolation of DNA.

Sonic treatment is also carried out in the cold at 5°C. Saline-EDTA washed cells are suspended in 10 vol saline-EDTA and placed in the cup of the sonic apparatus (Raytheon). The suspension is exposed to 9 kc (50 w) sound waves for 5 to 30 min. Sodium lauryl sulfate is then added, the suspension centrifuged to remove any cell debris and then subjected to the isolation procedure. If the DNA does not precipitate as threads upon ethyl alcohol addition, the nucleic acids are collected by centrifugation.

Lyophilization of Cells

Lyophilization is a commonly used laboratory technique for the preservation of cells. It was thought of interest to examine the DNA isolated from *D. pneumoniae* which had been subjected to freeze drying. The method employed for the isolation of DNA from the lyophilized cells is the same as that described using sodium lauryl sulfate as the lysing agent.

(f) Isolation of DNA by Cesium Chloride Density Gradient Centrifugation

The introduction of the cesium chloride density gradient technique by Meselson, Stahl & Vinograd (1957) has made it possible to separate RNA, DNA and protein under mild conditions. By selecting the proper density and conditions of centrifugation, the

RNA collects at the bottom of the centrifuge tube, the protein floats on the top and the DNA bands at or near the center. Biologically active DNA can readily be located by assaying the transforming activity of various fractions. The technique can also be adapted to separate DNA samples differing in base composition (and thus in buoyant density).

As applied to the isolation of DNA from cell lysates, the following outline has been used. A concentrated cell suspension of *D. pneumoniae* is lysed by sodium lauryl sulfate and the mixture shaken with a concentrated solution of CsCl containing 0·005 M-tris buffer (2-amino-2-hydroxymethylpropane-1:3-diol) plus 0·005 M-EDTA, pH 8·0. After adjusting the density to $1·706 \pm 0·002$ g/ml., the mixture is centrifuged in lusteroid tubes in the model L Spinco preparative centrifuge at 30,000 rev/min in the SW-39 rotor at room temperature for 3 days. When the rotor head has coasted (unbraked) to a stop, the tubes are removed, secured firmly in a vertical position and a small hole bored in the bottom of the tube with a small-gauge needle. The collected fractions, when sufficiently diluted to eliminate the inhibitory effects of CsCl on transformation (1:200 final dilution in the case of *D. pneumoniae*), are then assayed for biological activity. The fractions showing peak activity are combined, dialysed and treated with RNase to purify further the DNA. In one experiment, the DNA thus isolated accounted for 80% of the ultraviolet absorbing material at 260 mμ.

3. Results

A representative group of DNA samples isolated from bacteria by either lysozyme or detergent cell rupture is listed in Table 1. The sedimentation coefficient (at 20 μg/ml.) and T_m values have been determined and recorded. (The sedimentation coefficient was the same regardless of whether the centrifuge cells—regular or synthetic boundary—were filled slowly with a syringe or a wide-tipped pipette.) Even though the organisms listed vary widely taxonomically as well as in the base composition of their DNA, the purified DNA samples have approximately the same sedimentation coefficients with calculated molecular weights of the order of 8 to 12 mil-

Table 1.

Properties of representative DNA samples isolated from cells sensitive to lysozyme and/or sodium lauryl sulfate

Organism	$S_{20,w}$	T_m
Aerobacter aerogenes	22·4	94·0
Bacillus brevis	25·3	87·1
B. cereus	24·2	82·7
B. megaterium	29·0	85·2
B. subtilis	26·0	87·4
Diplococcus pneumoniae	24·0	85·2
Escherichia coli (K12)	29·0	90·0
Hemophilus influenzae	28·5	85·5
Klebsiella pneumoniae	25·6	92·0
Micrococcus lysodeikticus	24·0	98·6
M. pyogenes var *aureus*	28·0	83·2
Pseudomonas aeruginosa	28·2	96·7
Salmonella typhimurium	24·3	91·1
Serratia marcescens	23·7	93·6
Shigella dysenteriae	22·7	90·0
Streptococcus salivarius	24·4	85·2
Streptomyces albus	25·0	100·1
Calf thymus	22·1	86·0
Salmon sperm	23·2	87·0

The sedimentation coefficient and T_m were determined using saline-citrate (0·15 M-NaCl plus 0·015 M-Na citrate) as the solvent. The T_m for *S. albus* was obtained at a lower ionic strength and corrected to that for saline-citrate solvent. The sedimentation coefficients, determined immediately or within a day after the DNA preparation, did not alter on prolonged storage.

lion. Calf thymus and salmon sperm DNAs, isolated and supplied by Dr. N. Simmons, are also listed for comparison. In determining the T_m values, the temperature-absorbance curves showed that all the DNA samples isolated were predominantly in the native configuration. Variations of the sedimentation coefficient and the T_m of different preparations of DNA from the same organism were of the order of ± 1 s and $\pm 0·3°C$, respectively. An example of the sedimentation pattern of DNA isolated by the present method is shown in Plate I.

The effects of mechanical disruption and lyophilization of cells on the transforming activity and molecular weight of *D. pneumoniae* DNA are compared to detergent lysis in Table 2. Cells that have been subjected to freeze drying yield DNA with no detectable differences from that isolated from freshly harvested cells. However, as might be expected, sonic treatment (Litt, Marmur, Ephrussi-Taylor & Doty, 1958) has a deleterious effect on DNA. Grinding with glass is less destructive than sonic treatment and

would thus be the preferred method of mechanical disruption. Other methods applying a shear force (Davison, 1959; Hershey & Burgi, 1960; Cavalieri & Rosenberg, 1959) would probably result in degradation of the DNA.

The result of the cesium chloride density gradient centrifugation technique for the isolation of DNA from *D. pneumoniae* is shown in Fig. 1. The fractions, assayed for transformation with respect to streptomycin resistance, exhibit a peak activity at a density of CsCl expected from the base composition of the DNA (Sucoka, Marmur & Doty, 1959). The pooled active fractions, freed of contaminating RNA as described previously, had a high specific transforming activity and a molecular weight of 7×10^6 (estimated from the sedimentation coefficient). It is possible that some shear degradation of the DNA took place during mixing the lysate with CsCl, fraction collection through the narrow orifice or syringing the purified DNA into the sedimentation velocity cell (Davison, 1959). Similar experiments carried out by Mr. C. Schildkraut and Mr. R. Rownd (unpublished) employing milder conditions have yielded DNA from *E. coli* with slightly higher sedimentation coefficients, in fact very similar to that for DNA isolated by deproteinization and alcohol precipitation as described. If links do exist between DNA molecules, they are disrupted by even the mildest conditions. Since the buoyant densities of the DNA in CsCl isolated by both methods are identical, one can conclude that both methods yield a product which contains the same amount of protein.

Several DNA samples were used to see whether they would transform homologous strains with respect to the streptomycin marker. Preliminary studies have shown that *E. coli* (K12), *E. freundii, M. pyogenes* var *aureus, Alcaligenes fecalis* and *Ps. aeruginosa* are incapable (less than 1 in 10^7) of being transformed, using techniques similar to those used for *D. pneumoniae* and *B. subtilis*.

4. Discussion

Micro-organisms possess DNAs that vary widely in base composition and thus offer an ideal source of this nucleic acid in the study of differences imposed

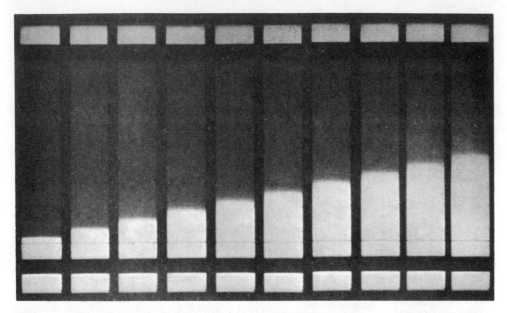

Plate 1. Ultracentrifuge sedimentation pattern of *E. freundii* DNA, 0.15 M NaCl + 0.015 M Na citrate. Using ultra-violet optics, pictures were taken every 4 min at 35,600 rev/min. First exposure at left. DNA concentration = 20 μg/ml.: 30 mm cell with Kel-F centerpiece. The sedimentation coefficient $S_{20,w}$ = 29.4.

Fig. 1. CsCl density gradient isolation of DNA from streptomycin resistant *D. pneumoniae*. The ability to transform to streptomycin resistance is plotted as a function of the sample obtained by collecting drops from the punctured, lusteroid tube after centrifugation.

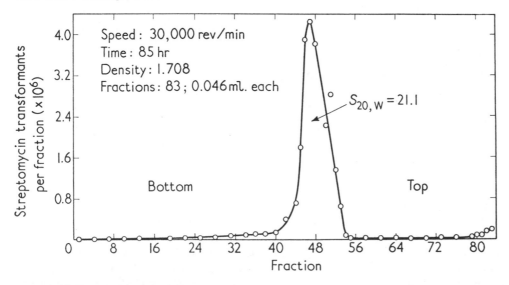

Table 2.

Effect of method of cell disruption and lyophilization on the transforming activity and molecular weight of *D. pneumoniae* DNA

Method of cell disruption	Relative transforming activity	Molecular weight $\times 10^{-6}$†
SLS‡	100%§	9·6
Grind with glass	36	5·0
Sonic 5′	4·1	1·0
Sonic 30′	0·42	0·38
Lyophilize, SLS	108	10·5

†Estimated from the sedimentation coefficient.

‡SLS = sodium lauryl sulfate.

§100% represents $3·6 \times 10^6$ transformants to streptomycin resistance per μg transforming factor.

on the molecule when the base pairs adenine:thymine and guanine:cytosine are present in different proportions. By studying the differences in the properties of DNA from different species, it is possible to gain an insight into the heterogeneity of the molecules comprising the genome of an organism.

The method described for the isolation of DNA has anticipated most of the difficulties that may be encountered with various micro-organisms. In general, the method described has yielded DNA from a variety of micro-organisms which is native, highly polymerized and possesses a fairly uniform molecular weight distribution. It might be argued that the method used for the isolation of DNA resulted in a degradation of the molecules and thus does not represent the true *in vivo* value. Degradation could have taken place during the shaking (Hershey & Burgi, 1960) used for deproteinization. The initial lysate is much more viscous than the final product; if the viscosity is due to a continuous DNA structure without protein links, then degradation has most likely taken place. The genetic evidence does indicate that the genome of *E. coli* behaves as one linkage group (Jacob & Wollman, 1958); however definitive evidence is still lacking as to whether this "chromosome" consists of large subunits or molecules (Forro & Wertheimer, 1960). Kellenberger (1960) has described a model in which protein "linkers" hold the DNA molecules together in the bacterial nucleoid.

The present method yields DNA from micro-organisms with a molecular weight of about 8 to 12 $\times$ 10^6 which is perfectly adequate for most experimental purposes. Those DNAs that possess transforming activity (e.g. *D. pneumoniae, B. subtilis,* etc.) can transform their homologous strains with an efficiency of 1 to 5% at saturating levels of DNA.

DNA isolated by selective buoyancy in a CsCl density gradient has physical, chemical and biological properties similar to the product obtained by the method described which involves selective denaturation with chloroform and alcohol precipitations. However, the DNA is still subjected to mild shear forces during the CsCl method and it is again possible that degradation may have taken place.

The methods of DNA isolation described in the literature, and here, are continually being improved and undoubtedly agents will be uncovered that will selectively precipitate or fractionate DNA under milder conditions than heretofore employed. The method presented in this report has been useful in providing samples of DNA from micro-organisms with widely varying base ratios and has proved of great value in studying their physico-chemical and genetic properties.

The author wishes to thank Dr. N. Simmons for valuable suggestions and discussions concerning the DNA isolation procedure and Dr. R. D. Hotchkiss who contributed greatly by initiating and instructing the author about various facets of DNA isolation and its biological activities. The author also wishes to thank Dr. P. M. Doty for his interpretations, criticisms and advice and Miss D. Lane who carried out the biological assays. Mr. C. Schildkraut contributed greatly with helpful discussions. Mr. J. Kucera and Mr. W. Torrey carried out some of the sedimentations and helped in the CsCl preparative DNA isolation.

References

Bayliss, M. (1937). *J. Lab. Clin. Med.* **22,** 700.

Bolle, A. & Kellenberger, E. (1958). *Schweiz, Z. Path. Bakt.* **21,** 714.

Cavalieri, L. F. & Rosenberg, B. H. (1959). *J. Amer. Chem. Soc.* **81,** 5136.

Chargaff, E. (1955). In *The Nucleic Acids,* ed. by E. Chargaff & J. N. Davidson, p. 308. New York: Academic Press.

Davison, P. F. (1959). *Proc. Nat. Acad. Sci., Wash.* **45,** 1560.

Doty, P., Boedtker, H., Fresco, J. R., Haselkorn, R. & Litt, M. (1959). *Proc. Nat. Acad. Sci., Wash.* **45,** 482.

Doty, P., McGill, B. B. & Rice, S. A. (1958). *Proc. Nat. Acad. Sci., Wash.* **44,** 432.

Eddy, A. A. & Williamson, D. H. (1957). *Nature,* **179,** 1252.

Forro, F. & Wertheimer, S. A. (1960). *Biochim. biophys. Acta,* **40,** 9.

Fox, M. S. & Hotchkiss, R. D. (1957). *Nature,* **179,** 1322.

Gunsalus, I. C. (1957). In *Methods of Enzymology,* ed. by S. P. Colowick & N. O. Kaplan, Vol. 1, p. 51. New York: Academic Press.

Hershey, A. D. & Burgi, E. (1960). *J. Mol. Biol.* **2,** 143.

Hotchkiss, R. D. (1957). In *Methods in Enzymology,* ed. by S. P. Colowick & N. O. Kaplan, Vol. 3, p. 692. New York: Academic Press.

Jacob, F. & Wollman, E. L. (1958). *Symp. Soc. Exp. Biol.* **12,** 75.

Kellenberger, E. (1960). *Symp. Soc. Gen. Microbiol.* **10,** 39.

Kirby, K. S. (1957). *Biochim. J.* **66,** 495.

Kirby, K. S. (1959). *Biochim. biophys. Acta,* **36,** 117.

Lerman, L. S. & Tolmach, L. J. (1957). *Biochim. biophys. Acta,* **26,** 68.

Litt, M., Marmur, J., Ephrussi-Taylor, H. & Doty, P. (1958). *Proc. Nat. Acad. Sci., Wash.* **44,** 144.

Marmur, J. & Doty, P. (1959). *Nature,* **183,** 1427.

McCarty, M. (1952). *J. Exp. Med.* **96,** 555.

Meselson, M., Stahl, F. W. & Vinograd, J. (1957). *Proc. Nat. Acad. Sci., Wash.* **42,** 581.

Sevag, M. G., Lackman, D. B. & Smolens, J. (1938). *J. Biol. Chem.* **124,** 425.

Sohler, A., Romano, A. H. & Nickerson, W. J. (1958). *J. Bact.* **75,** 283.

Spackman, D. H., Stein, W. H. & Moore, S. (1958). *Analyt. Chem.* **30,** 1190.

Spizizen, J. (1959). *Fed. Proc.* **18,** 957.

Sueoka, N., Marmur, J. & Doty, P. (1959). *Nature,* **183,** 1429.

Weibull, C. (1958). *Ann. Rev. Microbiol.* **12,** 1.

Zamenhof, S., Reiner, B., DeGiovanni, R. & Rich, K. (1956). *J. Biol. Chem.* **219,** 165.

Nucleic Acids II. Characteristics of *E. coli* DNA

Introduction
The following exercises are designed to demonstrate your effectiveness in isolating a high yield of native *E. coli* DNA. Each calls for the use of the UV spectrophotometer and provides data regarding the concentration of DNA, the helicity of DNA, and the level of protein or RNA contamination in the sample. Briefly, you are to determine the absorption spectrum of the DNA sample before and after alkaline denaturation as well as after overnight alkaline hydrolysis. The operation of the UV spectrophotometer is quite similar to that of the colorimeter, but you should get specific directions for its use from the laboratory instructor.

Procedure
1. Absorption spectrum of native DNA
   a. Prepare a sample of DNA dissolved in 3.0 ml of standard saline citrate (0.15 *M* NaCl, 0.015 *M* trisodium citrate, pH 7.0) and transfer it to a quartz cuvette. Prepare a 3.0-ml reagent blank of standard saline citrate (SSC) and transfer it to a matched quartz cuvette.
   b. Set the wavelength selector of the spectrophotometer at 260 nm and determine the absorption of the DNA sample at that wavelength. If the reading is in excess of 1.0, prepare a more dilute sample. The dilute sample should have a volume of 3.0 ml and an absorbance at 260 nm of between 0.5 and 0.7. Record the dilution factor.
   c. Determine the absorption spectrum at 10-nm intervals from 230 nm to 300 nm.
2. Absorption spectrum for denatured DNA
   a. Using the blank and sample from step 1, add 0.33 ml of 1.0 *M* KOH to each solution directly in the cuvettes. Cover the cuvettes with parafilm, invert several times to mix thoroughly, and determine the absorption spectrum again.
   b. The addition of KOH should bring the final concentration to 0.1 *M* KOH, which is sufficient to denature the DNA fully.
3. Alkaline hydrolysis to approximate RNA content of the sample
   a. Prepare a 5.0-ml sample of DNA dissolved in SSC having an absorbance at 260 nm of between 0.3 and 0.7. Prepare a blank tube containing 5.0 ml of SSC.
   b. Add 0.55 ml of 5.0 *M* KOH to each tube, mix, and determine the absorbance at 260 nm.
   c. Cover each of the tubes with parafilm and place in a 37°C water bath overnight.
   d. Chill the tubes in an ice bath, add to each several drops of concentrated perchloric acid (70% $HClO_4$), and mix. Check with pH paper to be certain that the solutions are acidic. This procedure will precipitate the denatured DNA but leave the hydrolysed RNA in solution.
   e. Leave the samples on ice for 10 minutes. Centrifuge at 4000 rpm for 10 minutes to bring down the precipitated DNA.
   f. Carefully remove the clear supernatant solution and determine its absorbance at 260 nm.

Calculations and Questions
1. Native DNA
   a. Plot the absorption spectrum of native DNA and examine the peak(s). A sharp peak at 260 nm indicates pure DNA; a broad peak with absorption near 280 nm indicates protein contamination.
   b. Calculate the concentration of DNA using the following relationship: a solution of DNA with an A_{260} of 1.0 contains 45 μg of DNA per milliliter.
   c. What is the total yield of DNA, and how does it compare with that predicted in the Marmur paper?
   d. Calculate the ratio of the absorbance at 260 nm to the absorbance at 280 nm. This ratio provides a rough measure of protein contamination. A DNA sample free of protein has a 260:280 ratio of about 2. Owing to the absorbance of protein at 280 nm, those samples containing protein have ratios less than 2.
2. Denatured DNA
   a. Plot the absorption spectrum of denatured DNA on the same graph paper used to plot the native DNA absorption.
   b. Calculate the percent hyperchromicity (h_{260}) using the following formula:

$$h_{260} = \frac{A_{260} \text{ denatured} - A_{260} \text{ native}}{A_{260} \text{ native}} \times 100$$

c. Highly polymerized DNA exhibits a hyperchromic shift of 30 to 40 per cent.

3. Alkaline hydrolysis

a. The relative amount of RNA in the sample is determined by comparing the A_{260} of the sample before and after the overnight hydrolysis procedure. The absorption reading before hydrolysis includes both DNA and any contaminating RNA that may be present. The absorption reading of the supernatant after hydrolysis and acid precipitation includes RNA only; therefore the difference between these two readings represents the total DNA content.

b. Calculate the relative amounts of DNA and RNA in the sample and express them in terms of percentages using the following relationship:

$$A_{260} \text{ before} - A_{260} \text{ after} = \text{DNA}$$
$$(\text{DNA} + \text{RNA}) \quad (\text{RNA})$$

4. Based upon your calculations, evaluate your sample of DNA as to:

a. Yield
b. Double-strandedness
c. Protein contamination
d. RNA contamination

5. What factors other than the presence of denatured DNA might contribute to a low hyperchromicity?

Additional Reading

F. B. Armstrong, *Biochemistry*, 2nd ed., Chaps. 13, 21, and 22. Oxford University Press, New York, 1983.

E. Chargaff and J. M. Davidson, *The Nucleic Acids*, Vols. 1–3. Academic Press, New York, 1955, 1960.

E. E. Conn and P. K. Stumpf, *Outlines of Biochemistry*, 4th ed. Wiley, New York, 1976.

J. N. Davidson, *The Biochemistry of the Nucleic Acids*. Methuen, London, 1965.

J. N. Davidson and R. M. S. Smellie, "Phosphorus Compounds in the Cell. 2: The Separation by Ionophoresis on Paper of the Constituent Nucleotides of Ribonucleic Acid," *Biochem. J.* **52**:594 (1952).

R. D. Hotchkiss, "Methods for the Characterization of Nucleic Acids," in *Methods in Enzymology*, S. P. Colowick and N. O. Kaplan, eds., Vol. III, pp. 708–715. Academic Press, New York, 1957.

A. L. Lehninger, *Biochemistry*, 2nd ed. Worth, New York, 1975.

J. Marmur and P. Doty, "Determination of the Base Composition of Deoxyribonucleic Acid from its Thermal Denaturation Temperature." *J. Mol. Biol.* **5**:109 (1962).

J. Marmur, R. Rownd, and C. L. Schildkraut, "Denaturation and Renaturation of Deoxyribonucleic Acid," in *Progress in Nucleic Acid Research and Molecular Biology*, J. N. Davidson and W. E. Cohn, eds., Vol. 1, pp. 232–300. Academic Press, New York, 1963.

APPENDIX

The following instructions are for the person or persons responsible for preparing the solutions and materials for each of the laboratory exercises. The amounts called for are predicated on a population of ten students working together as pairs (five laboratory groups). Solution volumes are higher than necessary to account for wastage in the laboratory. For larger class sizes, you must increase the amounts proportionately.

It is advisable to look over the requirements for each chapter about two weeks in advance so that missing items can be located or ordered and so that you can estimate how long the preparation will take.

In addition to the specific materials required for each laboratory exercise, the preparator will need some standard equipment and materials. Although the following list is not exhaustive, it does provide an idea of what should be available to set up each of the laboratories.

Assorted glassware (test tubes, graduated cylinders, beakers, erlenmeyer flasks, volumetric flasks, funnels, etc.)

Pipettes of various sizes (0.1 through 10.0 ml)

Propipettes

Pasteur pipettes and rubber bulbs

Bottles with caps of assorted sizes and types (glass and plastic)

Pipette soaker, washer, and dryer

pH meter and standard buffer for calibration

pH paper (0 to 13 range)

Parafilm

Scissors

Stirring motor and magnetic stir bars

Label tape and indelible markers

Balances (triple-beam and analytical)

Aluminum foil

Hot plate

Temperature-controlled water bath

Refrigerator, freezer

Fume hood

Safety glasses or goggles

Heat-resistant gloves

Thermometers (Celsius)

Filter paper

Distilled or deionized water

Chapter 1. Basic Techniques: The Preparation of Aqueous Solutions for the Laboratory

I. Solution preparation
   A. Reagents
      1. Sodium acetate—200 g
      2. Sodium hydroxide—100 g
      3. Concentrated hydrochloric acid—100 ml (this should be kept in a fume hood)
      4. Glacial (concentrated) acetic acid—100 ml (this should be kept in a fume hood)
      5. Distilled water—50 liters

B. Equipment and materials
1. Erlenmeyer flasks (or beakers) — 1,000 ml (20), 500 ml (5)
2. Volumetric flasks — 1,000 ml (5), 250 ml (10)
3. Pipettes — 10.0 ml (20)
4. Propipettes (10)
5. Triple beam balances (5) and weighing paper
6. Spatulas (5)
7. Labeling tape (5 rolls) and marking pens (5)
8. Pipette soaker

II. Measurement of pH
A. Reagents
1. Solutions prepared by students in first exercise
2. Distilled water — the 50 liters specified above is sufficient for all exercises
B. Equipment and materials
1. pH meters (2)
2. Standard buffer solution (pH 7.0) to adjust pH meter
3. pH paper — 0 to 13 range (5 rolls)
4. Pasteur pipettes (50) and rubber bulbs (30)
5. Beakers — 20 to 30 ml (25)
6. Wash bottles containing distilled water (5)

III. Buffering capacity
A. Reagents
1. Solutions prepared by students in first exercise
2. Distilled water — as above
B. Equipment and materials
1. pH paper — 0 to 13 range (5 rolls)
2. Pasteur pipettes (50) and rubber bulbs (30)
3. Pipettes — 10.0 ml (30)
4. Erlenmeyer flasks — 50 ml (10)
5. Pipette soaker

IV. pH and buffering capacity of some natural solutions

The reagents and materials used in parts I, II, and III of this exercise should be adequate for doing these independent determinations. The laboratory instructor should provide or suggest several natural liquids, such as vinegar, blood plasma, leaf exudate, etc.

Chapter 2. Colorimetry:
A Spectrophotometric Analysis of Riboflavin

I. The absorption spectrum of riboflavin
A. Reagents
1. Riboflavin solution (5.31×10^{-5} M) — carefully weigh out 0.02 g of riboflavin (M.W. = 376.37) and dissolve in distilled water to a final volume of 1 liter (use volumetric flask). If the riboflavin does not go into solution readily, you may need to heat it gently in a water bath until dissolution occurs. Store this solution refrigerated in a labeled bottle until about 1 hour prior to the laboratory period.
2. Distilled water — 1 liter
B. Equipment and materials
1. Colorimeters (5) equipped with clean, dry colorimeter tubes
2. Test tube racks (5)
3. Clean, dry test tubes — 16 $\times$ 150 mm (200)
4. Pipettes — 1.0 ml (50), 5.0 ml (50), 10.0 ml (50)
5. Propipettes (10)
6. Pasteur pipettes (200) and rubber bulbs (30)
7. Metric rulers or calipers (5)
8. Pipette soaker

II. A standard curve for riboflavin

The reagents and equipment described for part I of this exercise are sufficient for part II, with the following addition.

Prepare the following "unknown" solutions for student determinations:

5.31×10^{-5} M Riboflavin (ml)	Distilled water (ml)	Concentration (M)
0.5	9.5	0.27×10^{-5}
1.5	8.5	0.80×10^{-5}
2.5	7.5	1.33×10^{-5}
3.5	6.5	1.86×10^{-5}
4.5	5.5	2.39×10^{-5}
5.5	4.5	2.92×10^{-5}

These solutions should be prepared prior to the laboratory period and each given a letter code. The actual concentration of the unknown should be revealed to the students only after they have made their determinations.

Chapter 3. Proteins:
The Identification of a Dipeptide
A. Reagents
   1. Dipeptides: purchase at least 10 mg each of five different dipeptides from a biochemical supply company. It helps to choose those that include fairly dissimilar amino acids to assure clear separation upon chromatography. Prior to the laboratory period, conceal the identity of each bottle by covering the label with a piece of white tape and mark it in accordance with a predetermined letter code.
   2. Hydrochloric acid (6.0 *M*)—carefully add 49.8 ml of concentrated HCl to 30 ml of distilled water and mix. This will generate some heat and should be performed in a fume hood while wearing safety glasses. When the solution has cooled to room temperature, add distilled water to a final volume of 100 ml. Store in a glass-stoppered bottle.
   3. Amino acid standards (1.0 mg/ml in 10% isopropyl alcohol)—dissolve 1.0 mg of each of the following amino acids in 1.0 ml of 10% isopropyl alcohol. It may be necessary to add a small amount of HCl to get some of the amino acids into solution. Store the solutions frozen in labeled test tubes covered with parafilm.

a. isoleucine	k. tyrosine
b. glutamic acid	l. serine
c. proline	m. methionine
d. lysine	n. alanine
e. leucine	o. glutamine
f. arginine	p. cysteine
g. valine	q. asparagine
h. glycine	r. aspartic acid
i. tryptophan	s. threonine
j. histidine	t. phenylalanine

   4. Distilled water—1 liter
   5. Ninhydrin reagent—dissolve 0.75 g of ninhydrin in 250 ml of *n*-butanol and add 7.5 ml of glacial acetic acid.
   6. *n*-butanol—1 liter
   7. Formic acid—1 liter
   8. Phenol saturated with water (Solvent II)—gradually add about 220 ml of distilled water to each of two 1-pound bottles of crystalline phenol. Gentle warming and shaking will hasten the dissolution of the phenol. Transfer the contents of these two bottles to a large separatory funnel and let stand until there is obvious separation of the aqueous (upper) and nonaqueous phases. Draw the lower phase (phenol saturated with water) off into a brown bottle and store in the hood.
   9. Ammonium hydroxide (0.3% v:v)—add 10.0 ml of concentrated NH_4OH (29%) to 990 ml of distilled water. Note that concentrated NH_4OH is also available at a concentration of 58%, in which case the dilution should be 200-fold rather than 100-fold.
   10. Ethyl alcohol (95%)—have two or three wash-bottles full of ethanol in the laboratory in case of phenol spills.
B. Equipment and materials
   1. Analytical balance, weighing paper, and clean spatulas
   2. Watch glasses—5-cm diameter (5)
   3. Capillary tubes—1.6 to 1.8 × 100 mm (50)
   4. Bunsen burners (5) and flint igniters
   5. Labeling tape (5 rolls) and marking pens (5)
   6. Oven (set at 100°C)
   7. Chromatography paper, Whatman No. 1—a sufficient number of large sheets to prepare twenty 21 × 26 cm sheets
   8. Additional equipment for preparing chromatography papers: shears, pencils, centimeter rulers (5 of each)
   9. Disposable plastic gloves (10 pairs)
   10. Stapler
   11. Paper towels
   12. Pipettes—0.1 ml (10)
   13. Micropipettes—20 μl (200) or capillaries (1.6 to 1.8 × 100mm) calibrated as described in the text.

14. Graduated cylinders—50 ml (5), 100 ml (5), 250 ml (5)
15. Heat guns or hair dryers (5)
16. Small triangular files (5)
17. Beakers—50 ml (10)
18. Fume hood with sufficient space to accommodate 20 chromatography jars and fitted with strings and clips from which to hang the chromatograms.
19. Atomizer or aerosol sprayer to spray chromatograms with locating solution.
20. Chromatography jars with lids (20) (Jars should be large enough to easily accommodate a cylindrical chromatogram 21 cm tall and 8 cm in diameter. Large battery jars covered with a watch glass or plate are ideal, but beakers or large mayonnaise jars can also be used.)
21. Pipette soaker

Chapter 4. Proteins: The Extraction and Purification of Wheat Germ Acid Phosphatase

I. Extraction procedure

A. Reagents

1. Cold distilled water (4°C)—3 liters.
2. Manganese chloride (1.0 M)—dissolve 20.78 g of $MnCl_2 \cdot 4\ H_2O$ in distilled water to a final volume of 100 ml.
3. Ammonium sulfate (saturated, pH 5.5)—dissolve 785 g of $(NH_4)_2SO_4$ in 1,100 ml of distilled water. Solution is hastened by warming the solution on a hot plate. This should yield about 2,000 ml of the saturated solution. Store the solution in the refrigerator.
4. Disodium ethylenediaminetetraacetic acid (0.2 M)—dissolve 16.3 g of Na_2H_2EDTA in distilled water to a final volume of 250 ml.
5. Methanol—1 liter stored in freezer (at −20°C) for 10 hours prior to use.
6. Disodium ethylenediaminetetraacetic acid (5.0 mM)—dilute 125 ml of 0.2 M

Na_2H_2EDTA (see above) to a final volume of 5 liters.

B. Equipment and materials

1. Wheat germ (fresh, not toasted), may be obtained at local grocery or food specialty stores (300 g)
2. Beakers—50 ml (5), 250 ml (5), 500 ml (10)
3. Ice buckets with crushed ice (5)
4. A refrigerated high-speed centrifuge with a minimum top speed of 10,000 × g. The centrifuge should be equipped with two rotors, the larger one capable of centrifuging loads in excess of 200 ml and the smaller rotor with a 50-ml capacity. Clean, dry centrifuge tubes (40) and centrifuge bottles (30) should also be available.
5. Two-pan balance—situated near the centrifuge
6. Graduated cylinders—25 ml (5), 500 ml (5), 1,000 ml (5)
7. Magnetic stirring motors (5) plus 5 stir bars
8. Pipettes—1.0 ml (20), 5.0 ml (20), 10.0 ml (20)
9. Clean, dry test tubes—16 × 150 mm (100)
10. Temperature-controlled hot water bath adjusted to 65 to 70°C
11. Thermometers—Celsius (5)
12. Glass rods with rounded ends—approximately 0.5 × 15 cm (5)
13. Dialysis tubing—1.3 cm diameter (200 cm)
14. Scissors—clean, stainless steel (5)
15. Pasteur pipettes (200) and rubber bulbs (20)
16. Erlenmeyer flasks—1,000 ml (5)
17. Parafilm (1 box)
18. A cold room or large refrigerator should be available for storing solutions and the dialysis procedure.
19. A freezer should be available for storage of methanol and enzyme fractions.
20. Test tube racks (10)
21. Pipette soaker

II. Protein assay (biuret method)
   A. Reagents
      1. Distilled water — 1 liter
      2. Bovine serum albumin (1.0 mg/ml) — dissolve 1.0 g of bovine serum albumin (Sigma, Fraction V) in about 80 ml of distilled water and dilute to a final volume of 100 ml. The BSA should be stored frozen in labeled containers and thawed only when needed for standard curve assays.
      3. Biuret reagent — place 1.50 g of copper sulfate ($CuSO_4 \cdot 5\ H_2O$) and 6.0 g of sodium potassium tartrate ($NaKC_4O_6 \cdot 4\ H_2O$) in a dry 1-liter volumetric flask. Add 500 ml of distilled water and dissolve the contents with constant swirling. Add, with constant mixing, 300 ml of 10% (w:v) NaOH. Add enough distilled water to give a final volume of 1 liter and mix thoroughly. The Biuret reagent should be a deep blue color and can be stored in a plastic bottle at room temperature indefinitely. If a black or red precipitate forms, discard the solution and prepare a fresh batch.
   B. Equipment and materials
      1. Clean, dry test tubes — 16 $\times$ 150 mm (200)
      2. Pipettes — 1.0 ml (50), 5.0 ml (50), 0.1 ml (50)
      3. Colorimeters (5) with clean, dry colorimeter tubes
      4. Test tube racks (5)
      5. Pipette soaker
III. Acid phosphatase assay
   A. Reagents
      1. Distilled water — 1 liter
      2. Sodium acetate buffer (1.0 M, pH 5.7) — carefully dissolve 5.74 ml of glacial acetic acid in 80 ml of distilled water. Check the pH and adjust it to 5.7 by adding small volumes of 10.0 M NaOH (40 g of NaOH dissolved in 100 ml of distilled water). Once the pH has been adjusted, add enough distilled water to give a final volume of 100 ml.
      3. Magnesium chloride (0.1 M) — dissolve 2.03 g of $MgCl_2 \cdot 6\ H_2O$ in distilled water to a final volume of 100 ml.
      4. p-Nitrophenyl phosphate (0.05 M) — dissolve 1.68 g of disodium p-nitrophenyl phosphate in distilled water to a final volume of 100 ml. This solution should be prepared on the day it will be used.
      5. Potassium hydroxide (0.5 M) — dissolve 28.1 g of KOH in distilled water to a volume of 500 ml.
   B. Equipment and materials
      1. Colorimeters (5) with clean, dry colorimeter tubes
      2. Parafilm (1 box)
      3. Stopwatches (5)
      4. Pipettes — 1.0 ml (50), 5.0 ml (50), 0.5 ml (50), 0.1 ml (50)
      5. Clean, dry test tubes — 16 $\times$ 150 mm (200)
      6. Test tube racks (5)
      7. Temperature-controlled water bath adjusted to 37°C.
      8. Desktop clinical centrifuge (1) with 15-ml-capacity centrifuge tubes (50)
      9. Pipette soaker

Chapter 5. Proteins: The Kinetic Properties of Wheat Germ Acid Phosphatase

In order to complete all five of the exercises, each group of students should have at its disposal approximately 30 ml of wheat germ acid phosphatase with sufficient activity to promote an absorbance change at 405 nm (ΔA_{405}) of approximately 0.35 using the standard acid phosphatase assay. The enzyme isolated during the previous exercise should be plenty, but if necessary the enzyme is commercially available and can be prepared at the proper dilution before the laboratory period.

The makeup instructions for most of the reagents required for this series of exercises can be found in the previous section and will be so designated with an asterisk (*). The recommended amount of each solution has been calculated on the assumption that each of the five groups will repeat each exercise three times.

I. Time course of the reaction
  A. Reagents
    1. Distilled water—500 ml
    2. Sodium acetate buffer* (1.0 *M*, pH 5.7)—75 ml
    3. Magnesium chloride* (0.1 *M*)—75 ml
    4. *p*-Nitrophenyl phosphate* (0.05 *M)*—75 ml
    5. Potassium hydroxide* (0.5 *M*)—300 ml
    6. Wheat germ acid phosphatase—5 ml/group
  B. Equipment and materials
    1. Colorimeters (5) with clean, dry colorimeter tubes
    2. Parafilm (1 box)
    3. Stopwatches (5)
    4. Pipettes—0.5 ml (15), 1.0 ml (45), 5.0 ml (30)
    5. Clean, dry test tubes—16 × 150 mm (150)
    6. Test tube racks (5)
    7. Temperature-controlled water bath adjusted to 37°C.
    8. Desktop clinical centrifuge (1) with 15-ml-capacity centrifuge tubes (50)
    9. Pipette soaker

II. The effect of different substrate concentrations on reaction velocity
  A. Reagents
    1. Distilled water—500 ml
    2. Sodium acetate buffer* (1.0 *M*, pH 5.7)—100 ml
    3. Magnesium chloride* (0.1 *M*)—75 ml
    4. *p*-Nitrophenyl phosphate* (0.05 *M*)—75 ml
    5. Potassium hydroxide* (0.5 *M*)—400 ml
    6. Wheat germ acid phosphatase—5 ml/group
  B. Equipment and materials
    1. Colorimeters (5) with clean, dry colorimeter tubes
    2. Parafilm (1 box)
    3. Stopwatches (5)
    4. Pipettes—0.1 ml (5), 0.5 ml (15), 1.0 ml (50), 5.0 ml (40)
    5. Clean, dry test tubes—16 × 150 mm (200)

6. Pipette soaker
7. Test tube racks (5)
8. Temperature-controlled water bath adjusted to 37°C
9. Desktop clinical centrifuge (1) with 15-ml-capacity centrifuge tubes (50)

III. The inhibition of acid phosphatase by inorganic phosphate
  A. Reagents
    1. Distilled water—500 ml
    2. Sodium acetate buffer* (1.0 *M*, pH 5.7)—100 ml
    3. Magnesium chloride* (0.1 *M*)—100 ml
    4. Potassium hydroxide* (0.5 *M*)—400 ml
    5. Potassium phosphate, dibasic (0.005 *M*)—dissolve 0.174 g of K_2HPO_4 in about 150 ml of distilled water and make up to a final volume of 200 ml
    6. Wheat germ acid phosphatase—5 ml/group
  B. Equipment and materials
    1. Colorimeters (5) with clean, dry colorimeter tubes
    2. Parafilm (1 box)
    3. Stopwatches (5)
    4. Pipettes—0.5 ml (15), 1.0 ml (60), 5.0 ml (30)
    5. Clean, dry test tubes—16 × 150 mm (200)
    6. Pipette soaker
    7. Test tube racks (5)
    8. Temperature-controlled water bath adjusted to 37°C
    9. Desktop clinical centrifuge (1) with 15-ml-capacity centrifuge tubes (50)

IV. The effect of different enzyme concentrations on reaction velocity
  A. Reagents
    1. Distilled water—500 ml
    2. Sodium acetate buffer* (1.0 *M*, pH 5.7)—100 ml
    3. Magnesium chloride* (0.1 *M*)—100 ml
    4. *p*-Nitrophenyl phosphate* (0.05 *M*)—100 ml
    5. Potassium hydroxide* (0.5 *M*)—500 ml
    6. Bovine serum albumin (1 mg/ml)—dissolve 0.1 g of BSA (Sigma, Fraction V) in

about 80 ml of distilled water and make up to a final volume of 100 ml. This solution should be stored in the freezer.

 7. Wheat germ acid phosphatase—5 ml/group

B. Equipment and materials

 1. Colorimeters (5) with clean, dry colorimeter tubes

 2. Parafilm (1 box)

 3. Stopwatches (5)

 4. Pipettes—0.1 ml (5), 0.5 ml (15), 1.0 ml (55), 5.0 ml (30)

 5. Clean, dry test tubes—16 × 150 mm (200)

 6. Pipette soaker

 7. Test tube racks (5)

 8. Temperature-controlled water bath adjusted to 37°C

 9. Desktop clinical centrifuge (1) with 15-ml-capacity centrifuge tubes (50)

V. Effects of different temperatures on reaction velocity

A. Reagents

 1. Distilled water—1,000 ml

 2. Sodium acetate buffer* (1.0 M, pH 5.7)—200 ml

 3. Magnesium chloride* (0.1 M)—200 ml

 4. p-Nitrophenyl phosphate* (0.05 M)—200 ml

 5. Potassium hydroxide* (0.5 M)—1,000 ml

 6. Wheat germ acid phosphatase—10 ml/group

B. Equipment and materials

 1. Colorimeters (5) and clean, dry colorimeter tubes

 2. Parafilm (1 box)

 3. Stopwatches (5)

 4. Pipettes—0.5 ml (30), 1.0 ml (130), 5.0 ml (60)

 5. Clean, dry test tubes—16 × 150 mm (400)

 6. Test tube racks (5)

 7. Temperature-controlled water baths (3) adjusted to 30, 37, and 50°C

 8. Ice buckets with crushed ice (5)

 9. Large beakers—400 to 600 ml (5)

10. Boiling water baths (2)—500-ml beaker of water on a hot plate

11. Desktop clinical centrifuge (1) with 15-ml-capacity centrifuge tubes (50)

12. Thermometers—Celsius (20)

13. Pipette soaker

Chapter 6. Carbohydrates: The Analysis of Glycogen

I. Glucose analysis

A. Reagents

 1. Glucose (dextrose) (0.5 mM)—carefully weigh out 0.09 g of glucose and dissolve in distilled water to a final volume of 1,000 ml (volumetric flask).

 2. Nelson's reagent A—dissolve 12.5 g of anhydrous Na_2CO_3, 12.5 g of potassium sodium tartrate, 10.0 g of $NaHCO_3$, and 100 g of anhydrous Na_2SO_4 in a final volume of 500 ml of distilled water. Leave the solution at room temperature and if a sediment forms remove it by filtration.

 3. Nelson's reagent B—dissolve 15.0 g of $CuSO_4 \cdot 5\ H_2O$ in 90 ml of distilled water containing 2 drops of concentrated H_2SO_4. Make up to a final volume of 100 ml with distilled water.

 4. Arsenomolybdate reagent—dissolve 25.0 g of ammonium molybdate in 450 ml of distilled water. Add to this solution 21.0 ml of concentrated sulfuric acid (do this slowly and carefully, with stirring). Dissolve 3.0 g of sodium arsenate ($Na_2HAsO_4 \cdot 7\ H_2O$) in 25.0 ml of distilled water and add this solution to the acid molybdate solution with stirring. Incubate the resulting solution at 37°C for 24 hours. Store the arsenomolybdate reagent in a brown bottle. There should be no green tint in this solution. Do not let it come into contact with aluminum, which is often found in the caps of bottles.

B. Equipment and materials

 1. Pipettes—1.0 ml (50), 2.0 ml (50), 5.0 ml (50), 10.0 ml (50)

2. Clean, dry test tubes — 16 × 150 mm (200)
3. Test tube racks (5)
4. Aluminum foil (1 box)
5. Boiling water baths (5) — 500 ml beaker of water on a hot plate
6. Colorimeters (5) with clean, dry colorimeter tubes
7. Graduated cylinders — 50 ml (5), 100 ml (5)
8. Stopwatches (5)
9. Propipettes (10)
10. Pipette soaker

II. Acid hydrolysis of glycogen
   A. Reagents
      1. The reagents prepared for glucose analysis above (Nelson's reagent A, Nelson's reagent B, and the arsenomolybdate reagent) are sufficient for this exercise.
      2. Hydrochloric acid (4.0 *M*) — carefully (while wearing safety glasses) add 33.2 ml of concentrated HCl to an equal amount of distilled water. This is an exergonic reaction, and the solution will heat up. When the solution has cooled, add enough distilled water to give a final volume of 100 ml.
      3. Glycogen (8 mg/ml) — dissolve 0.8 g of glycogen in a final volume of 100 ml of distilled water.
      4. Potassium phosphate dibasic (1.0 *M*) — dissolve 174.2 g of K_2HPO_4 in distilled water to a final volume of 1,000 ml.
   B. Equipment and materials — identical to those needed for glucose determination

III. Analysis of glycogen by thin layer chromatography
   A. Reagents
      1. Hydrochloric acid (4.0 *M*) — the amount of 4.0 *M* HCl prepared above is sufficient for both exercises.
      2. Glycogen (100 mg/ml) — add 10 g of glycogen to 100 ml of distilled water and mix. It may be necessary to heat this solution in a water bath to get the glycogen into solution. Filter out any residual sediment.
      3. Potassium phosphate, dibasic (1.0 *M*) —

the amount of 1.0 *M* K_2HPO_4 prepared above is sufficient for both exercises.
      4. Sodium hydroxide (1.0 *M*) — dissolve 8 g of NaOH in distilled water to a final volume of 200 ml.
      5. Hydrochloric acid (1.0 *M*) — exercising the same cautions described above, mix 16.6 ml of concentrated HCl with 16.6 ml of distilled water. After it has cooled, make up to a final volume of 200 ml with distilled water.
      6. Potassium phosphate buffer (0.5 *M*, pH 6.9) — dissolve 17.24 g of KH_2PO_4 and 21.77 g of K_2HPO_4 in 200 ml of distilled water. Adjust the pH to 6.9 by the addition of concentrated NaOH or HCl. Dilute to a final volume of 250 ml with distilled water.
      7. Sodium chloride (0.1 *M*) — dissolve 1.17 g of NaCl in distilled water to a final volume of 200 ml.
      8. Potassium phosphate, monobasic (0.1 *M*) — dissolve 13.6 g of KH_2PO_4 in distilled water to a final volume of 1,000 ml.
      9. Glucose (1.0% [w:v] solution in 10% [v:v] propanol) — prepare 100 ml of 10% propanol by dissolving 10 ml of propanol (propyl alcohol) in distilled water to a final volume of 100 ml. Dissolve 1.0 g of glucose in 100 ml of 10% propanol.
      10. Maltose (1.0% [w:v] solution in 10% [v:v] propanol) — prepare in the same manner as glucose, above.
      11. Chromatography solvent (chloroform: glacial acetic acid:water, 30:35:5) — mix 300 ml of chloroform with 350 ml of glacial acetic acid and 50 ml of distilled water.
      12. Locating solution (1.23% [w:v] *p*-anisidine and 1.66% [w:v] phthalic acid in methanol) — dissolve 2.46 g of *p*-anisidine and 3.32 g of phthalic acid in 200 ml of methanol (methyl alcohol).
   B. Equipment and materials
      1. Clean, dry test tubes — 16 × 150 mm (100)
      2. Test tube racks (5)

3. Pipettes—1.0 ml (20), 2.0 ml (20), 5.0 ml (20), 10.0 ml (20)
4. Boiling water baths (5)—500-ml beaker of water on a hot plate
5. Aluminum foil (1 box)
6. Thin layer chromatography plates—silica gel on glass or plastic backing (5 × 20 cm); Baker-Flex Silica Gel IB2-F works well (20)
7. Shallow glass dish (at least 10 × 25 × 2 cm); a lasagna baking dish is ideal (2)
8. Oven set at 110°C
9. Pencils (5), metric rulers (5)
10. Micropipettes—10 μl (100)
11. Heat guns or hair dryers (5)
12. Chromatography jars (5)—at least 25 cm tall and 8 to 10 cm in diameter
13. Large watch glasses to cover chromatography jars (5)
14. Atomizer or aerosol sprayer to spray chromatograms with locating solution
15. pH paper—0 to 13 range (5 rolls)
16. Pasteur pipettes (200) and rubber bulbs (30)
17. Beakers—50 ml (10)
18. Fume hood for spraying and drying the TLC plates
19. Ice buckets (5) containing crushed ice
20. Pipette soaker

Chapter 7. Lipids:
An Analysis of Some Common Fats and Oils

CAUTION: No flames in the laboratory—since several highly volatile and explosive solvents are used in these exercises, there should be no open flames in the laboratory at any time.

I. Saponification
  A. Reagents
    1. Lipid samples, all of which can be obtained at a grocery store
      a. Lard—100 g
      b. Soybean oil—100 g (alternative vegetable oils which may be used include corn oil and olive oil)
      c. Butter—10 g
      d. Margarine—10 g
    2. Ethyl alcohol (95%)—500 ml
    3. Sodium hydroxide (50%, [w:v])—dissolve 50 g of NaOH pellets in 80 ml of distilled water. Handle with care; this solution will generate considerable heat. Add sufficient distilled water to give a final volume of 100 ml.
    4. Distilled water—400 ml
  B. Equipment and materials
    1. Triple-beam balances (3)
    2. Small watch glasses for weighing out lipids—5 to 10 cm diameter (10)
    3. Erlenmeyer flasks—250 ml (5)
    4. One-hole rubber stopper to fit the mouth of the 250-ml flask; fit the stopper with a length of glass tubing (120 cm × 0.5 cm) as shown in Figure 7-13 (5)
    5. Hot plates (5)
    6. Ring stands with two clamps (5)
    7. Beakers—500 ml (5), 1,000 ml (5)
    8. Safety glasses or protective goggles (10)
    9. Ice buckets with crushed ice (5)
    10. Small metal spatulas, approximately 1 × 15 cm (5)
    11. Magnetic stirring motors and stir bars (5)
    12. pH paper—0 to 13 range (5 rolls)
    13. Pasteur pipettes (200) and rubber bulbs (30)
    14. Test tube racks (5)
    15. Clean, dry test tubes—16 × 150 mm (50)
    16. Pipettes—5.0 ml (10), 10.0 ml (20)
    17. Graduated cylinders—100 ml (5), 500 ml (5)
    18. Pipette soaker
II. Solubility of lipids
  A. Reagents
    1. Lard, soybean oil, butter, and margarine (the amounts called for above are sufficient for both exercises)
    2. Distilled water—200 ml
    3. Chloroform—200 ml, in the hood
    4. Diethyl ether—200 ml, in the hood
    5. Ethyl alcohol—200 ml

B. Equipment and materials
1. Triple-beam balances and watch glasses, as above
2. Clean, dry test tubes — 16 × 150 mm (200)
3. Test tube racks (5)
4. Spatulas — as above (5)
5. Pipettes — 5.0 ml (40)
6. Celsius thermometers (5)
7. Pasteur pipettes (200) and rubber bulbs (30)
8. Pipette soaker

III. Fatty acid isolation
A. Reagents
1. Hydrochloric acid (concentrated) — 100 ml in a glass-stoppered bottle
2. Distilled water — 1 liter
3. Soap solutions prepared in part I

B. Equipment and materials
1. Pasteur pipettes (200) and rubber bulbs (30)
2. pH paper — 0 to 13 range (5 rolls)
3. Celsius thermometers (5)
4. Glass stirring rods — approximately 25 × 0.5 cm (5)
5. Beakers — 250 ml (10)
6. Spatulas (5)
7. Hot plates (5)
8. Ice buckets plus crushed ice (5)
9. Triple-beam balances and watch glasses, as above
10. Small brown bottles with screw-top closures (10)
11. Pipettes — 5.0 ml (20)
12. Graduated cylinders — 100 ml (5)
13. Pipette soaker

IV. Thin layer chromatography
A. Reagents
1. Lipid samples
a. Fatty acid samples isolated in part III
b. Soybean oil and lard
2. Petroleum ether (B.P. 35 to 60°C) — 1,000 ml, in the hood
3. Silver nitrate (5% [w:v] in 50% methyl alcohol [v:v]) — dissolve 5.0 g of $AgNO_3$ in 50 ml of distilled water and add 50 ml of methyl alcohol. This solution is light sensitive and should be stored in a brown bottle in the dark.
4. Acetic acid (glacial) — 25 ml, in a glass-stoppered bottle
5. Sulfuric acid (50% [v:v]) — in a fume hood and while wearing safety glasses and gloves, *carefully* pour 50 ml of concentrated H_2SO_4 into a flask containing 50 ml of distilled water. As the acid reacts with the water, a great deal of heat will be generated and the flask will become hot. This solution should be stored in a glass-stoppered bottle.
6. Diethyl ether — 100 ml (leave in its original container in the fume hood)

B. Equipment and materials
1. Triple-beam balances and watch glasses, as above
2. Small brown bottles with screw-top closures (20)
3. Pipettes — 1.0 ml (20), 10.0 ml (25)
4. Graduated cylinders — 100 ml (5)
5. Thin layer chromatography plates — silica gel on glass or plastic backing (5 × 20 cm), Baker-Flex Silica Gel IB2-F works well but becomes brittle in the sulfuric acid locating solution (30)
6. Shallow glass dish (10 × 25 × 2 cm) for dipping TLC plates (5)
7. Paper towels
8. Forceps (10)
9. Oven — set first at 100°C for activation of the TLC plates and later at 180°C for localizing the spots on the plates
10. Pencils (5)
11. Centimeter rulers (5)
12. Micropipettes — 5 μl (50)
13. Fume hood with enough bench space for 5 chromatography jars and strung with lines and clips on which to hang the chromatograms
14. Covered chromatography jars lined with filter paper — jars should be large enough to accommodate two 5 × 20 cm TLC plates (5)

15. Enamel trays—30 × 45 cm (3)
16. Pipette soaker

Chapter 8. Nucleic Acids: Isolation and Characterization of *E. coli* DNA

I. Isolation of DNA from *E. coli* cells
  A. Reagents
    1. *E. coli* cells, 2 to 3 g of wet-packed cells for each pair of students. In reality, practically any bacterium readily available in large quantities can be used in this exercise.
      a. Lyophilized *E. coli* cells specifically for DNA extraction are available from Sigma Chemical Co. Just prior to the laboratory period, suspend 5 g of lyophilized cells in 500 ml of saline-EDTA (see below). Centrifuge for 10 minutes at 5,000 × *g* to pellet the cells, discard the supernatant, and distribute the bacterial paste uniformly among the student groups.
      b. *E. coli* or other bacterial cells may also be grown on standard bacteriological media and harvested just before the laboratory period.
    2. Saline-EDTA (0.15 *M* NaCl plus 0.1 *M* disodium ethylenediaminetetraacetic acid, pH 8.0)—dissolve 17.52 g of NaCl and 74.4 g of $Na_2EDTA \cdot 2 H_2O$ in 1,900 ml of distilled water. Adjust the pH to 8.0 by the dropwise addition of concentrated NaOH to the solution. Add distilled water up to a final volume of 2,000 ml.
    3. Sodium lauryl sulfate (25%)—dissolve 25 g of sodium lauryl sulfate (sodium dodecyl sulfate) in 80 ml of distilled water. Make up to a final volume of 100 ml with distilled water. This should be made up at least 1 day in advance because the SLS dissolves slowly.
    4. Lysozyme, crystalline—bottle containing at least 100 mg.
    5. Sodium perchlorate (5.0 *M*)—dissolve 306 g of $NaClO_4$ (anhydrous) in 400 ml of distilled water. Make up to a final volume of 500 ml with distilled water.
    6. Chloroform-isoamyl alcohol (24:1, v:v)—mix 1,920 ml of chloroform with 80 ml of isoamyl alcohol.
    7. Ethyl alcohol (95%)—200 ml
    8. Three concentrations of saline citrate
      a. Concentrated saline citrate (CSC): 1.5 *M* NaCl plus 0.15 *M* trisodium citrate, pH 7.0—dissolve 43.85 g of NaCl and 22.0 g of trisodium citrate · 2 H_2O in 450 ml of distilled water. Adjust the pH to 7.0 and make up to a final volume of 500 ml with distilled water.
      b. Dilute saline citrate (DSC): 0.015 *M* NaCl plus 0.0015 *M* trisodium citrate, pH 7.0—pipette 10.0 ml of CSC into a 1,000-ml volumetric flask and dilute to volume.
      c. Standard saline citrate (SSC): 0.15 *M* NaCl plus 0.015 *M* trisodium citrate, pH 7.0—pipette 100 ml of CSC into a 1,000-ml volumetric flask and dilute to volume.
    9. Ribonuclease (0.2% [w:v] in 0.15 *M* NaCl, pH 5.0)—prepare a solution of 0.15 *M* NaCl (0.88 g of NaCl dissolved in distilled water to a final volume of 100 ml) and adjust the pH to 5.0. Dissolve 0.01 g of ribonuclease in 5.0 ml of the NaCl solution. Place the ribonuclease solution in an 80°C water bath for 10 minutes to inactivate deoxyribonucleases. Cover the tube with parafilm and store in the freezer.
    10. Acetate-EDTA (3.0 *M* sodium acetate plus 0.001 *M* EDTA, pH 7.0)—dissolve 40.8 g of sodium acetate and 0.034 g of disodium EDTA in 80 ml of distilled water. Adjust the pH to 7.0. Make up to a final volume of 100 ml with distilled water.
    11. Isopropyl alcohol—100 ml
  B. Equipment and materials
    1. A refrigerated, high-speed centrifuge with a minimum top speed of 13,000 × *g*. The centrifuge should be equipped with

large and small rotors, plus tubes, adapters, and bottles so that it can handle loads from 20 to 200 ml.

2. Temperature-controlled water baths (2) adjusted to 37°C and 70 to 80°C

3. Erlenmeyer flasks with ground-glass stoppers—250 ml (5)

4. Beakers—50 ml (10), 100 ml (10), 500 ml (10)

5. Pipettes—1.0 ml (30), 2.0 ml (30), 5.0 ml (30), 10.0 ml (30)

6. Pasteur pipettes (200) and rubber bulbs (30)

7. Clean, dry test tubes—16 × 150 mm (500)

8. Test tube racks (5)

9. Stirring motor, fitted with a glass stirring rod with a screw taper; used to stir the solution (500 to 1,000 rpm) during the isopropanol addition

10. Graduated cylinders—100 ml (5)

11. Glass stirring rods—approximately 0.5 × 25 cm (10)

12. Pipette soaker

II. Characterization of *E. coli* DNA

A. Reagents

1. Dilute saline citrate, standard saline citrate, and concentrated saline citrate; the amounts prepared for the DNA extraction should suffice.

2. Potassium hydroxide (1.0 *M*)—dissolve 5.61 g of KOH in distilled water to a final volume of 100 ml.

3. Potassium hydroxide (5.0 *M*)—dissolve 28.06 g of KOH in distilled water to a final volume of 100 ml.

4. Perchloric acid (70%)—10 ml

B. Equipment and materials

1. UV spectrophotometers (2) with matched quartz cuvettes (4)

2. Pasteur pipettes (200) and rubber bulbs (30)

3. Pipettes—1.0 ml (30), 2.0 ml (30), 5.0 ml (30), 10.0 ml (30)

4. Clean, dry test tubes—16 × 150 mm (100)

5. Test tube racks (5)

6. High-speed centrifuge as described for the DNA extraction

7. Temperature-controlled water bath adjusted to 37°C

8. pH paper—0 to 13 range (5 rolls)

9. Ice buckets (5) and crushed ice

10. Pipette soaker

INDEX